FIGHTING ROBOTS

FIGHTING ROBOTS

A GUIDE TO RADIO-CONTROLLED COMBATANTS

MICHAEL BENSON

CITADEL PRESS
Kensington Publishing Corp.
www.kensingtonbooks.com

CITADEL PRESS BOOKS are published by

Kensington Publishing Corp.
850 Third Avenue
New York, NY 10022

All Kensington titles, imprints, and distributed lines are available at special quantity discounts for bulk purchases for sales promotions, premiums, fund-raising, educational, or institutional use. Special book excerpts or customized printings can also be created to fit specific needs. For details, write or phone the office of the Kensington special sales manager: Kensington Publishing Corp., 850 Third Avenue, New York, NY 10022, attn: Special Sales Department, phone 1-800-221-2647.

Designed by Leonard Telesca

First printing: July 2002

10 9 8 7 6 5 4 3 2 1

Printed in the United States of America

Library of Congress Control Number: 2002100673

ISBN: 0-8065-2372-7

WARNING: Building a fighting robot can be very dangerous. Do not build a robot or operate one unless you are qualified to do so—or are supervised by a qualified adult.

Contents

Acknowledgments

The author wishes to gratefully acknowledge the following individuals and organizations: John Adipietro; Fred Barton, America's Robot Man; Mark Beiro, the world's greatest ring announcer; Tekla Benson; Jeff Cesnik of Cesnik Technologies; Dan Creagan; Jake Elwell, literary agent extraordinaire; Chris Hannold at Litewav Robotics; Skip Harmeyer and the Sylvan Lake Robot Warriors; Mark Joerger; Peter Julian at Robo '02; Walter Martinez and the Robotics Society of Southern California; Kenneth Maxon; Chuck Pitzer at Paleo-Robotic Research Laboratories; Seattle Robotics Society; and Noel Sharkey of Creative Robotics at Magna.

Preface: Bots to the Rescue

Not all of the newly developed technology regarding radio-controlled robots is going into making TV shows. Bots are being used to help mankind in real life, often taking on dangerous or time-consuming duties that formerly had to be handled by human beings.

Recently, in war-torn Bosnia, a robot set a record for recovering unexploded ordnance—a task for which, for obvious reasons, it is very difficult to find human volunteers.

Bots at Ground Zero

Robots were also used following the Twin Towers disaster to search for survivors beneath the mammoth piles of rubble. The radio-controlled bots were equipped with TV cameras that sent continuous images back to the human driver.

The bots sent back signals so that the driver would always know precisely where they were. They sent pitch-and-roll information so that the driver would know if they were going uphill or downhill, or perhaps running along the side of an incline. They had several temperature sensors, and even had thermal "night vision" capabilities, so that they could search in the dark and distinguish between a living and dead body.

These bots are built so that, like a Slinky in reverse, they can *climb* stairs. This enables them to search a multistoried building without ever needing a human present. One of the reasons that the bots have been so effective at the World Trade Center is that they are unaffected physically and emotionally by the horrendous working conditions. They are no more bothered by the dust and the smoke than they are by the unspeakable evil of the attack and its resulting carnage.

The World Trade Center disaster has offered the bots, with their lights, video cameras, and sensors, one of their first opportunities to partake in a human search-and-rescue mission—although, sadly, in this case, there were no survivors to find.

Mini Battle Tanks

The bots that were used at Ground Zero ranged in size from that of a shoe box to that of a medium-sized suitcase. Some have compared their appearance to mini battle tanks, as the rescue bots are built exclusively with caterpillar-treaded tires.

The bots belong to both the United States military and federal disaster units. They are prototypes financed by the National Science Foundation and the Office of Naval Research.

The bots were originally designed for tasks as varied as military reconnaissance, defusing explosives, and safety tests at nuclear and chemical installations. Although it is frightening to think about, the robots would work equally well at the scene of a nuclear, biological, or chemical attack. Some had to be brought out of retirement after the September 11 attacks.

"Everyone that physically saw them was impressed by the technology," said Lieutenant Colonel John Blitch, an army robotics expert who coordinated the teams, in a *New York Times* interview. "The fire department loved them."

On the afternoon of September 12, a small robot called a MicroTrac crawler located a pocket of rooms that had remained largely intact under a pile of debris. Emergency workers then uncovered the rooms and a firefighter was sent down. Several bodies were found.

Robots also located the remains of other victims and surveyed three damaged buildings near the World Trade Center site. Some of the World Trade Center robots were small and compact enough to go down sewage pipes less than a foot in diameter.

Others could flatten themselves to squeeze through tight spaces or position themselves upright to look over objects. Some of the robots were developed or customized through the Tactical Mobile Robots program of the

Defense Advanced Research Projects Agency, or DARPA, an arm of the military that finances technological research and development.

Colonel Blitch

The Tactical Mobile Robots program was, in part, the creation of the agency's former director, Lieutenant Colonel Blitch, who took part in rescue efforts after the bombing of the Oklahoma City federal building in 1995.

Blitch saw how difficult it was for humans to maneuver safely through the debris, and he became convinced of the need for small, agile robots.

"I saw a lot of situations where we had a hard time penetrating into narrow void spaces and conduits," Blitch said.

Blitch was, at the time, a University of South Florida graduate student in Tampa. Blitch's adviser there was Robin Murphy, an associate professor in computer science and engineering who does robotics research.

Blitch returned to school to tell his adviser that there was a humanitarian need for robots. Since then, both have increasingly focused on rescue and recovery robots—Blitch through the DARPA program and Dr. Murphy at the university.

Blitch, who was planning to move to Colorado to lead the National Institute for Urban Search and Rescue's new Center for Robot-Assisted Search and Rescue, still had robots stored in the basement of his house in Washington on September 11, 2001.

After the two planes hit the World Trade Center, he packed the robots into his car and called other robotics teams as he drove to New York. With all nonmilitary air traffic grounded, representatives from Foster-Miller and iRobot, two other robotics companies, drove down from the Boston area.

Dr. Murphy and three graduate students drove eighteen hours from Tampa. The military's Space and Naval Warfare Systems Command brought in robots from San Diego. The robotics teams converged in Newburgh, New York, about sixty miles north of New York City.

Bot Spex

The prices of the rescue robots start at $15,000 to $30,000, and those price tags can climb much higher depending on the number and types of sensors and other features they have, including two-way audio and night vision.

The small MicroTrac crawler and the briefcase-size Micro Unmanned Ground Vehicle, which were brought by Lieutenant Colonel Blitch, are customized versions of robots made by Inuktun, a company based in British Columbia. Foster-Miller had two robots at work: the Solem, which is about the size of a shoe box, and the Talon, which has arms and pincers.

The Packbot, manufactured by iRobot, features flippers that allow it to climb stairs and hills. Urbots, which were brought by the Space and Naval Warfare Systems Command, can function upside down.

Military officials have ordered more robots from Foster-Miller and Inuktun. This early deployment of the robots has alerted researchers to features that are needed but not yet developed: for one, temperature sensors to be used when penetrating burning rubble. Dr. Murphy said that a robot that was sent into the depths of the rubble lost its rubber treads, probably because they were melted by the fires smoldering under the debris.

And so, I dedicate this book to Lieutenant Colonel John Blitch, Dr. Robin Murphy, and all of the men and women who have, step by step, developed the technology that, even as it entertains us with combat, helps to save lives under the harshest of all conditions.

Introduction: Mechanical Warriors

They look like sinister lawn mowers, and they are the warriors in what has become the fastest growing sport on TV: battling robots. Competitors from all over the world are building fighting machines in their attics, basements, and garages.

Like geek gladiators they gather in one spot for the meet. There, using radio-control technology, the inventors pit their mechanical warriors in fights to the death.

They do not hold this contest in just any arena. Even the playing field is dangerous. It is booby-trapped with buzz saws that come out of the floor and lined with potentially impaling spikes.

Scoring Points

Each robot scores points by ramming its opponent into various booby traps about the arena, by avoiding its opponent's blows through superior agility and speed, and by using its weapon, whatever it may be, to inflict damage upon its foe.

As in boxing, some bouts end in knockouts and others go the distance. The fights last a maximum of three minutes apiece. The winner, in that case, must be decided by a panel of judges.

Some robots come with hydraulic hammers that smash their opponents. Some have chain saws to slice the mechanical guts from their foes. Others are built like wedges to get under and flip over the opposition.

This guide to the world of metallic maulers was written by yours truly, the author of *Monster Trucks* and the former editor of the *Military Technical Journal*. Hopefully, it will appeal both to those who enjoy fighting robots because of the action and to the "techno-geeks" (and, of course, I mean that in only the nicest possible way) who want to build and fight their own machines.

The DIY Set

It has been said that those who construct battling robots are from the DIY sector of life—that is, the Do-It-Yourself set. These men and women originate from all walks of life.

The innovative inventors who make these competitions possible are machinists, computer programmers, rocket scientists, sculptors, and architects—yet they inevitably have one thing in common: They have a passion for understanding how things work.

Because that passion is so common, the fighting robot fever is spreading like wildfire. During the summer of 2001, bot builder Christian Carlberg was interviewed on NBC's *Today* show by host Matt Lauer, and Carlberg captured the fever in words: "I saw a competition about five years ago in San Francisco and I was hooked. I think anybody who watches this sport gets hooked immediately. It brings out the kid in you. You get to build these toys, you get to play with them, you get to break them, and you get to build them again. And it is good clean fun. You are breaking bolts, not bones."

At first glance, the sport of fighting robots appears to lack human drama—but, after watching for a while, you realize that nothing could be further from the truth. That's because behind each robot is a team of real human beings. The robot not only reflects its creator's ingenuity, but each bot's battling skills are dependent on its driver's ability to operate the remote control.

So here's to the botz! And to the men and women who build and drive them! Excelsior!

Michael Benson
Brooklyn, N.Y.
Autumn 2001

1

A History of Robotic Combat

Folks have been experimenting with robot competitions since as far back as 1980, when Mattel, the toy company, sponsored competitions between man-made machines. Machines that most quickly and efficiently performed some simple task were the winners.

But, if you had to pick one day to be the birthday for the sport of robot fighting, it would be the day in 1992 when Marc Thorpe, who had graduated twenty-one years earlier from the University of California at Davis, was trying to save himself (and perhaps the rest of the world) some labor by inventing a remote-controlled vacuum cleaner.

When the resulting machine tore up his living room rather than cleaned his rug, a lightbulb went on over Marc's head. He planned a meet of robot builders, with fights to the death—of the robots, that is.

First Event

Between the turning on of the lightbulb and the culmination of that idea, two years passed. The first "Robot Wars" was held at the Fort Mason Center in San Francisco, California.

At that first event—which was sponsored in part by a record label in New York City owned by Steve Plotnicki called Smile Communications—remote-controlled robots fought each other in front of a live audience within an arena.

The audience was not made up of a handful of people who had happened to wander in off the street. The first event was attended by an enthusiastic audience numbering close to a thousand.

Bot historians will be interested to know that the concept of booby-trapping the arena to make the battles more fun and dangerous was Thorpe's and was there in the competitions from the very beginning.

The second event, held a year later, also at the Fort Mason Center, sold out. Lessons had been learned during that first event. This one featured bots that were stronger and more dangerous than before.

Sue Me Sue You

The third annual Robot Wars, held in 1996 in the same arena, had its ups and downs. The event was a huge success—a record seventy-five bots attended and fought—but behind the scenes there were problems.

Thorpe and Steve Plotnicki, whose record company had changed its name to Profile Records in the preceding year, ended up going to court over who got how much of the proceeds.

Those courtroom troubles lasted a full year and threatened the 1997 event. The event went on, and was a huge success, only after a last-second agreement. But after that event, Steve Plotnicki decided to hold Robot Wars events on his own, without Thorpe as a partner.

Also in 1997, Plotnicki teamed up with a British television production company called Mentorn Broadcasting. Together they produced six episodes of a robot combat series, which aired over the BBC2 network over six weeks.

Following the event, Plotnicki and Thorpe got into a legal scrape over who was the rightful owner of the Robot Wars concept. Plotnicki tried to prevent Thorpe from putting on his 1998 show.

While this was going on, interest in the battling robots, expressed prolifically in Internet chat rooms, was sky-high. The focal point for the Internet robot chat was Robot Wars Forum, a Web site owned and operated by Carlo Bertocchini, who was also the builder of a popular battling bot known as BioHazard.

Although legal difficulties prevented the annual Robot Wars 1998 from being held at the Fort Mason Center, one of the competitors, a fellow by the name of Gary Cline, put together an alternative called Robotica (not to be confused with the Learning Channel television show starring Ahmet Zappa). The event was to be held at the Cow Palace in San Francisco on the same weekend that the Fort Mason event was originally scheduled, August 14–16.

Around this time, predominantly through the Robot Wars Forum Web site, a unified group of competitors known as the Society of Robotic Combat (SORC) was formed—sort of a union of combatants apart from the courtroom melee.

The Robotica event did not go off without a hitch, however. One month before it was to occur, Mentorn, the producers of the British six-episode television series, announced that they were going to hold a public competition, also in San Francisco and also on the weekend of August 14–16.

Simultaneous to this, Mentorn was also filming its second TV series for broadcast in England. (Mentorn's San Francisco event, as it turned out, never came off, because public support was so strong for the Robotica event.)

With only days left before the big weekend, Gary Cline had to cancel the Robotica event after encountering more legal difficulties. With many of the world's battling-robot builders in San Francisco with nothing to do, an emergency meeting of the SORC was called.

The meeting was addressed by Thorpe, who explained that he was losing the legal battle to continue with the sport he invented. Court battles continued for another year. In March of 1999 an announcement was posted on the Robot Wars Forum Web site by a fellow named Gary Pini.

Profile Records ended up winning the rights to the Robot Wars trademark and Mark Thorpe quietly disappeared from the scene. That same month, an announcement was made that was going to change the course of battling robot history.

Enter BattleBots

The builders of LaMachine, a robot that had fought at the San Francisco Robot Wars shows, announced that they were creating a new competition known as BattleBots. Those men were Edward "Trey" Roski III, whose interest in extreme sports started at an early age—he climbed Mount Everest and went scuba diving when he was thirteen—and his cousin, Greg Munson of Oakland.

Also in March, Carlo Bertochinni resigned as the director of the Robot Wars Forum Web site—which had been rendered inactive due to lawsuits—and was replaced by Dan Danknick and Tony Buchignani, both of whom were aligned with the British Robot Wars people.

In response to this change, the lawsuit against the Web site by Profile Records was dropped, unfortunately leaving Carlo Bertochinni with thousands of dollars in unpaid legal bills. A couple of fellow robot builders helped him out a bit with the bills, but for the most part he took the burden upon himself.

In April of 1999 the Web site again became active. There was an announcement of a 1999 Robot Wars event and for the first time the Web site featured streaming video of previous battling bot events.

Steve Plotnicki and Profile Records incorporated their efforts as Robot Wars LLC and sued the BattleBots people on April 27, 1999, for attempting to cause them "enormous economic harm" by scheduling a competing event.

On June 14, 1999, a court determined that the Robot Wars people had no right to stop the BattleBots show. The following month the 1999 Robot Wars event was canceled and the Robot Wars Forum closed down.

The Lights Are On

On August 14 and 15, 1999, BattleBots held its first event in Long Beach, California, and it was a huge success. BattleBots took the lead in fighting robot competitions, a lead that it has maintained ever since.

BattleBots held its fights in an arena called the BattleBox, which added to the thrills of the combat by providing its own traps and pitfalls for the robots to avoid, even as they were trying to engage each other.

The BattleBox hazards included:

1. The Kill Saws. These were buzz saws that rose up out of the floor. Push your opponent onto one of these babies and he would be sliced and diced. Each saw contained four twenty-inch, carbide-tipped, SystiMatic saw blades. Those blades were powered by a 220-volt, 3.5-horsepower motor and could rise and lower with more than sixty pounds of pneumatic force.
2. Pulverizers. These were heavy-duty aluminum mallets, strong enough to pound any bot into so much tenderized robot meat.
3. Ram Rods. These were six carbide-tipped spears that rose out of the floor with over sixty pounds of pneumatic force, enough to impale the strongest of bot armor. Each rod was an inch in diameter, and

had been reinforced with four steel gussets. The rods protruded six inches from the floor.

4. The Spike Strip. The outer edge of the BattleBox fighting area was lined with 180 carbide-tipped spikes. Push your foe "onto the ropes" in the BattleBox and these "ropes" bit back. Each one-inch-diameter spike was sharpened to a fine point that protruded three inches from the steel-reinforced back plate.
5. The Vortex. A series of turntables were built into the BattleBox floor. Drive onto one of them and pretty soon you're heading in the wrong direction. The Vortex was designed more to upset your battle plan than to damage your bot.
6. The Hell Raisers. These were sections of the floor that unexpectedly popped up. If they popped up when you were on top of one, your bot might suddenly resemble a belly-up turtle. Compressed air raised a hidden piece of one-eighth-inch steel plate six inches into the air. The Hell Raisers measured three feet by four feet.

Box Spex

The BattleBox was designed by Peter Lambertson. With only minor changes, the BattleBox used today remains the same. Here are the specifications:

The BattleBox floor is a collection of 144 modules. Each module is four feet by four feet by two feet. Each module has a steel superstructure covered by one-eighth-inch steel floor plating.

The floor is painted with black traction paint. All floor modules are laser leveled to match each other in terms of height. The floor is ferrous. The dimensions of the finished area are forty-eight feet by forty-eight feet by two feet.

The walls of the BattleBox are made out of Lexan and supported with steel square tubing. The walls are one inch thick and twenty-four feet high. There is a thick-grade plastic ceiling covering the BattleBox.

Robots enter the BattleBox via eight-foot-by-sixteen-foot ramps. There are two eight-foot-by-eight-foot entry and exit doors. Bots begin their match from either a red square or a blue square, which are located on opposite sides of the arena.

Contestants control their bots from an eight-foot-by-twelve-foot-by-three-foot stage—either the red stage or the blue stage.

Pay-per-view

The first event went so well that a second BattleBots event was scheduled immediately, this time at the All American Sports Park in Las Vegas, Nevada, on November 17, 1999.

It would turn out to be the most important event in the history of the sport because it was videotaped with professional production values and would later be shown as a pay-per-view TV event on January 29, 2000.

That pay-per-view gave the sport its biggest boost ever. The show, which earned an unexpectedly large buy rate, put battling robots on the map for keeps.

The winners of BattleBots competitions receive as much as $6,000 and a big trophy shaped like a nut. They also have their names engraved forever on The Giant Nut, which is the Stanley Cup of BattleBots.

Another fighting robot competition was held on the weekend of May 6 and 7, 2000, in Mesa, Arizona. The contest featured smaller-sized bots only and was put together with the full cooperation of the BattleBots people. The show even used one-quarter of the BattleBots set and a few members of the BattleBots staff showed up to lend a hand.

Comedy Central?

In May 2000, it was announced that BattleBots and the Comedy Central cable network had worked out a deal. It seemed like an odd match at first. Bots tearing each other to pieces were funny in a way, in an *Itchy and Scratchy* kind of way, but was it really comedy?

What we didn't get was that Comedy Central didn't give a hoot whether it was comedy or not. They only cared whether people would watch it—and they were pretty sure people would.

The next BattleBots event would be videotaped and shown on TV as a series to run at the end of the year. The event that turned into the TV show was the one held from June 9 to 11, 2000, in San Francisco.

The venue was a familiar one for the veteran bot builders. It was held at the Fort Mason Center, where the first Robot Wars show had been held in 1994. All four weight classes—superheavyweight, heavyweight, middleweight, and lightweight—participated.

"To England!"

In the meantime, the Robot Wars people abandoned the United States to the BattleBots folks and took their act to Great Britain. After inviting all bot builders to attend their show and compete, all travel expenses paid, they managed to recruit only seven robots. That show was eventually shown on MTV.

The *BattleBots* TV show debuted on Comedy Central on Wednesday, August 23—with each episode repeated twice on Saturdays. The show had personality. The host of the show was Bil (one *l*, honest) Dwyer, who had a stand-up comedy act when he wasn't commenting on bot battles.

The ring announcer for the show, the guy who actually introduced the bots before each fight, was Mark Beiro, the Floridian articulator who had, for years, been the ring announcer on the USA Network boxing series, *Tuesday Night Fights.*

The next BattleBots live show took place November 16 to 19, 2000, at the All American Sports Park in Las Vegas, Nevada.

The videotape from that show was edited into half-hour episodes for Comedy Central. A record 147 robots showed up for the event. For the first time ever there were so many robots there and ready to compete that not all of the fights were going to make it onto TV.

The Chin

On October 25, 2000, BattleBots received a publicity coup when builders Christian Carlberg and Lisa Winter appeared on NBC's late-night talk show, *The Tonight Show with Jay Leno.*

Leno was so impressed with their bots that he vowed to build one of his own. On November 3, BattleBots again appeared on *The Tonight Show.* (This time it was Scott LaValley with DooAll, Don Hutson with Diesector, and Morgan Tilford with Mauler.) On that show Leno announced that his own robot, named Chin-Killa (after Leno's own most noteworthy facial feature), was finished.

On November 9, one-quarter of the BattleBox was moved onto *The Tonight Show* stage and Chin-Killa faced off against a bot named Mauler 2000. If ever there was a heavy betting favorite, it was that Jay Leno's bot was going to come out on top in this one.

Chin-Killa also appeared at the November BattleBots show in Las Vegas, taking on a bot named Ginsu (like the knives). Once again, the power of the chin was too much for Chin-Killa's puny foes.

A Global Chess Game

The Robot Wars people, still busy in England, formed an alliance with Nintendo, who made a Robot Wars game for its Game Boy Color.

The second season of *BattleBots* on Comedy Central started on December 12, 2000. In January 2001, the first season of *BattleBots* was shown for the first time on the Canadian edition of Comedy Central.

In February 2001, *BattleBots* launched its own Web site. During that same month *BattleBots* made a guest appearance on the *Nash Bridges* TV show with Don Johnson and Cheech Marin.

BattleBots took its act abroad when the first season of the TV show began broadcasting over the BBC2 in England in March 2001.

Robotica: The TV Show

A month later, *BattleBots* got its first competition for market share in the United States since Robot Wars crossed the sea. A new show appeared on The Learning Channel (TLC, on cable), called *Robotica.*

Robotica's premise was different from that of *BattleBots.* Instead of immediately pitting each competitor against another in a fight to the death, robots had to demonstrate other skills.

Those skills were strength, speed, and agility. The robots that proved superior at those skills earned the right to combat an opponent, the "Fight to the Finish," in what was the climax of each episode.

Every episode featured four competitors, divided into two pairs. The pairs would first race on a figure-eight track called the Speedway, and were encouraged to smash into one another whenever they could, especially at the intersection.

Then they would run through a maze to test the radio-control operators' skills at the controls. The third challenge was called the Gauntlet, where each bot had to show its strength and go over difficult surfaces while pushing heavy objects out of the way.

***Robotica* CHALLENGES**

The Speedway

The first *Robotica* challenge was the Speedway, which was set up like a figure eight. The robots started at opposite ends of the track and raced in opposite directions. The idea was for the two opposing bots to confront each other as often as possible during the race.

Though the bots were not allowed to use their weapons during the first challenge, they were encouraged to bump and crash into their opponent as violently as they pleased. Some bots craved the contact more than others. After all, for some bots, smashing into ones opponent *was* their primary weapon.

The first robot to complete eight laps or to complete the most laps at the end of two minutes won the event. Each lap successfully completed earned 10 points. The winner was awarded 20 bonus points.

Use of the Speedway as a *Robotica* challenge was discontinued after the first *season*.

The Labyrinth

The second *Robotica* challenge was the Labyrinth. During the first season this challenge was known as the Maze. For this challenge, both bots started on side-by-side, identical maze tracks. They had to drive up a seesaw ramp, which lowered them back to the ground on the other side.

Then they pushed a big cube out of the way and turned a sharp corner. If they could survive the following encounter with a series of spiked, swinging pendulums, the bots next climbed up giant speed bumps, which led them to the guillotine.

If they could get through the big blade of the guillotine without getting chomped in half, the robots had to climb a slick wet ramp with cascading water. At the top of the ramp was a waterfall. The first to pass under the waterfall won.

The robots received 15 points for each obstacle successfully navigated, and a 25-point bonus went to the first bot to reach the center of the maze. The maze was revamped for the show's second season, and the name was changed to the Labyrinth.

Robots now navigated a maze in which they had to overcome

six deadly challenges. If these could be overcome, the bot got to smash six glass walls, chalking up points the whole way.

The six challenges were a twenty-pound box, a shaky suspension bridge, a section of lethal spikes that shot up from the floor, a hydraulic flip ramp that could send a two hundred-plus-pound robot flying, and finally, a sand pit and a set of steel cargo rollers.

If a bot busted all six glass walls, a door at the end of the track opened, revealing a seventh glass wall. The robot that broke this final wall ended the game and received bonus points.

The trip through the Labyrinth was further complicated by the *Robotica* "rats," robotic rodents armed with buzz saws. Bots that got past the sand and the rollers earned 20 points. Those that got past the flip ramp and the spikes got 15 points. The bot that could clear the box and suspension bridge got 10 points, and the bot that managed to break the final strip of glass got a bonus of 15 points.

The Gauntlet

The third *Robotica* challenge was the Gauntlet, which tested the bots' power, speed, and endurance. For the first season of the show, this was a race, side by side, through a series of increasingly difficult-to-budge barriers.

A sheet of glass, a stacked pyramid of paint cans filled with sand, a wall of bricks, and a wall of concrete blocks stood in the robots' way before they could get to the final obstacle, a one hundred-pound safe that had to be pushed across the finish line.

Each obstacle cleared was worth 15 points. If a robot was unable to break through one obstacle, it was not permitted to go around it and strike the next obstacle. The first robot to break through the final obstacle won and was awarded 25 bonus points.

There was a three-minute time limit. If neither robot completed the course in three minutes, the robot farthest along was declared the winner. The second season found a new gauntlet for bots to get through.

The Gauntlet changed to a baseball-diamond-shaped track in which bots were forced to crash through four increasingly obstinate walls: a wall of wood, followed by one of weighted cans, then bricks, and finally a cement wall, and that was just to get to second base.

When the robots got back "home" they turned and climbed a

ramp up to the "Central Killing Zone." Once inside they found a "forest of glass" columns that had to be destroyed. When they completed this task, a final ramp led them out of the center area and on to victory.

Scoring was done thusly: For each untouched obstacle passed, the robot received 10 points. For getting through other obstacles ("rubble"), the robot received 5 points. The first robot up the ramp to the forest of glass received 10 points. The second robot up the ramp received 5 points. The robot that destroyed the final pane of glass received 15 points.

The Fight to the Finish

If a robot did better than his opponent during the preliminary rounds, it got to move on to the final round, the last event of each television episode, the Fight to the Finish.

The robots began this event on bridges leading to an eight-foot-high platform that was surrounded by blazing torches. Once the bots were on the platform, the bridges were removed and the bots were left alone, facing one another on the elevated platform.

The platform had short guardrails to keep the bots from falling off while allowing the live and TV audience to see what was going on. At the sound of a buzzer the fight commenced.

Now the bots were allowed to use their weapons. After three minutes, the guardrails lowered, and the first bot to fall off lost. There was no scoring for this event. The last bot on the platform was the winner and moved on to the next stage of the competition.

The loser went home—after its owner picked it off the spikes that stuck up below the edges of the elevated platform.

The season started with twenty-four bots in the competition. Over six episodes, this was whittled down to six bots. Those six then went at it in the season finale to determine the overall winner.

The TV host for *Robotica* was the large, bespectacled, and bald Ahmet Zappa, son of Frank and brother of Dweezle, who added an element of humor with his flamboyant introductions.

Helping Zappa were two lovelies: Tanika Ray—former host of *The Big Help* on Nickelodeon and *Limozeno* on The Den—and former Miss World Canada (1993/94), Tanya Memme.

From May 22 to 28, 2001, *BattleBots* taped their third season at Treasure Island in San Francisco. Robot Wars proved they were still in the game by holding (and videotaping) a show in London.

Toys

In July 2001 *BattleBots* toys went on sale in stores across the country and the third season of battles began on Comedy Central. The toys are robotic replicas that allow fans to recreate the excitement of *BattleBots* in the comfort of their own homes.

Marketed under JAKKS' Pacific Road Champs brand, the officially licensed *BattleBots* line includes "Metal Mechanics," build-it-yourself model sets with battery-operated weapons features.

There is also the "BattleBox Arena Playset," complete with authentic pounding pulverizers, kill saws, and realistic sound effects. Other miniature robotic models include the "Battle Bashers," "Grip 'N Grapplers," and "Battle Chains." The *BattleBots* line is available through *BattleBots*' online store. Prices range from $6.00 to $60.00.

Robot Wars USA

On Monday, August 20, 2001, *Robot Wars* debuted on American television, appearing on the TNN cable network. The network correctly assumed that the product, filmed in England, would go over big with the crowd who enjoyed professional wrestling. Therefore they put the show on Monday nights as a lead-in to the WWF on The National Network (TNN, formerly The Nashville Network).

To further maximize the crossover potential of battling bots and battling behemoths, the U.S. version of *Robot Wars*—known as *Robot Wars: Extreme Warriors*—featured WWF wrestler (and best-selling author) Mick Foley (a.k.a. Cactus Jack, ManKind, Dude Love, et al.) in videotaped intros for the bouts.

TV Guide

One week later, the fighting robot phenomenon earned a story in *TV Guide* (August 25–31, 2001, issue) called "Crash Course." The article proclaimed the beginning of "Robotmania."

On August 28, 2001, *BattleBots* announced that they had signed a licensing agreement with Majesco, Inc., to develop and market software for the Nintendo Advance Game Boy Color system. The video games were expected to make their debut in 2002.

BattleBots Season 4

During the week of November 11, 2001, *BattleBots* taped their fourth season in San Francisco. For this season, there were a few changes in the BattleBox. It had been determined that the hazards and weapons in the box were affecting the outcome of matches, and this was not the original intention. Therefore the center of the arena was cleared of hazards.

The driver of a bot now knew that his machine was safe from everything except its opponent as long as it stayed in the center.

There was also a warning light, inside the arena near the starting lights, that was designed to alert drivers of hazards that were functioning. There was also an override button so that all hazards could be deactivated in case a match ended abruptly.

In the past, there had been a problem with BattleBox hazards doing damage to losing bots after the match was over—with the new override button, this would no longer be the case.

Also in November 2001 it was announced that the second season of *Robot Wars: Extreme Warriors,* the series shown on TNN in the United States, was going to tape in January 2002 in London, England.

In December 2001, the second season of *Robotica* debuted on The Learning Channel. The revamped show debuted new challenges that featured lots of crowd-pleasing breaking of glass.

Which pretty much brings us up to the present. The sport of robotic combat continues to grow larger every day, and there is no end to that growth in sight.

The Future

What is the future for battling robots? For one thing, I think there will be more and more competition for autonomous rather than radio-controlled bots. The R/C bots we are used to seeing on TV rely on their drivers, the ladies and gentlemen outside the arena who are working the bots' controls, to know when to attack and when to retreat.

In the future robots will have internal computers programmed to sense what is going on around them, and to sense what is happening to them, so they will not need to be told what to do.

These computerized bots will be able to react to danger (and to opportunities for attack) many times quicker than human beings can instruct them through radio control, thus making the sport much more exciting for spectators. Bot fights that in the past looked like fights between blind mice might now more resemble a squaring off of Bruce Lee and Jackie Chan.

Since it will still be human beings programming the bots, the human element will still be there.

2

A History of Hollywood Robots

Mankind (the species, not the pro wrestler) has been fascinated with the idea of mechanical men as far back as the nineteenth century, when the subject cropped up again and again in the works of writer E. T. A. Hoffman.

As far as we know, the first robot to appear in a film came in 1897 in a one-minute movie made by the French film innovator Georges Melies called *The Clown and the Automaton.* The word "robot" was first popularized in a 1923 play called *Rossum's Universal Robots* written by Karel Capek.

Silent Bot

The first classic robot to appear in motion pictures was called Maria, a beautiful female bot from the much-acclaimed1927 German silent movie *Metropolis,* directed by Fritz Lang.

In the film, which takes place in a big city in the then-futuristic year 2001, the villain, Rottwang, delights in making a woman robot that is indistinguishable from a real woman.

We get to see his creation in its raw metallic form. Of course, we never get to hear the bot speak because this was two years before "talkies." Later in the film, after the "flesh" is applied, the bot is played by Brigitte Helm.

In the bot's most memorable scene, it does a scandalous dance that sets a crowd of men into a frenzy.

Serial Botz

Robots became a staple of a form of movies popular in the early days of sound film, the serial. Pretty much wiped out by television in the early 1950s, serials were film shorts that preceded the main attractions.

Each segment was about fifteen minutes long and ended in a cliff-hanger that required attendance the following week to find out what happened. Many serial villains—best exemplified by Bela Lugosi, of *Dracula* fame, as the villain in a 1939 serial called *The Phantom Creeps*—sought to take over the world with the help of a grim-faced killer robot.

In the 1936 serial *Undersea Kingdom,* hero Ray "Crash" Corrigan must defeat a whole army of robots to escape imprisonment at the ocean's bottom.

A villain played by Eduardo Ciannelli built a tin robot with a pointy nose in his attempt to take over the world in the Republic serial *Mysterious Doctor Satan,* which came out in 1940.

In the *Monster and the Ape,* a 1945 serial from Columbia, the "Monster" was a robot.

The list goes on and on. In the world of movie serials, many an episode ended with a robot's "hands" wrapped around the hero's throat.

When you think about it, it shouldn't come as a shock that robots were portrayed as villains in the movies of the 1930s and 1940s. This was an era when there weren't enough jobs to go around. The robots that were being developed in real life were intended to perform specific tasks, like screwing the caps onto tubes of toothpaste, or putting the steering wheels onto Ford automobiles—tasks that had previously required human hands.

More robots, it was feared, meant even fewer jobs—and a whole lot less food on the dinner table.

Gort

With the coming of the 1950s and the atomic age, science fiction grew up quickly. It graduated from kiddie fare to social commentary faster than you can say Cold War. A robot played a key role in the 1951 science fiction classic *The Day the Earth Stood Still.*

The robot was called Gort, and although it had arms and legs, it disturbingly lacked a face. Gort emerges from a flying saucer that lands in Washington, D.C. The robot and his humanlike companion (played by

Michael Rennie) were on Earth to spread the now clichéd message: "Clean up your act or the Earth is doomed."

The key to getting Gort to work? The words *"Klaatu barada nikto."*

Gort was portrayed by a seven-foot actor named Lock Martin, whose previous claim to fame was as a doorman at Grauman's Chinese Theatre.

If you are interested in purchasing a plastic model of Gort, contact Sci-Fi Matters at 2599 Diamond Street, San Francisco, California 94131-3009, or call 415-469-0699, or fax 415-587-6029.

Tobor the Great

In *Tobor the Great,* a 1954 film directed by Lee Sholem and produced by Richard Goldstone out of Republic Studios, two brilliant scientists create a robot for the purpose of exploring deep space.

All is well until the mechanical marvel is stolen by enemy agents. Only the scientists' psychic link with the robot can save it from being reprogrammed for evil purposes.

When villainous spies kidnap a scientist and his family, the loyal robot he brought to life sets out to rescue them. Perhaps you have already noticed that "Tobor" is "robot" spelled backward.

Robby the Robot

If you took a poll, you would find that most people's favorite movie robot is the bubble-topped Robby the Robot, a character in the 1956 science-fiction movie *Forbidden Planet,* which was an update of Shakespeare's *The Tempest.* The movie was produced by Nicholas Nayfack and directed by Fred McLeod Wilcox.

The six-foot, eleven-inch Robby, like Gort before him, also lacked a face—but he had something the others had lacked, a voice. As the "servant" of the future, Robby was great to look at, with parts that turned around and lots of dials and switches. In the script he has been programmed to do anything except hurt human beings.

During the film we get to see Robby help a chef prepare a meal, walk, talk, and drive. Though there are those who say Robby sports the most personality in the entire film, he is used mostly for comic relief.

Robby was so popular that he reappeared in another film, *The Invisible*

Boy, also made by MGM, in 1957, also starring Philip Abbott and Richard Eyer. In a publicity shot for that film, Robby was shown playing baseball, bat in hand.

During this film Robby was not just played for laughs. In the script, an evil computer tries to take over the world. It almost succeeds and even manages to corrupt Robby for a time.

We get to see Robby do a lot more than play Julia Child in this movie. The eight hundred-pound bot shoots destructive rays, and demonstrates incredible strength as he lifts a battle tank.

For those of you who won't sleep tonight without knowing, Robby gets his "conscience mechanism" flicked near the end of the film and returns to the side of niceness and goodness. Whew.

"Danger, Will Robinson . . ."

The most popular robot in TV history is a nameless bot. They called him Robot, and since he was the only bot around there was never any confusion. The robot was, of course, from *Lost in Space.*

The series, created by Irwin Allen, was set in the then-futuristic year of 1997, and was an update of the classic *Swiss Family Robinson* story. It ran on CBS from September 15, 1965, through September 11, 1968.

In the show, the Robinson family are sent by rocket into space. Their assignment: to colonize a planet in Alpha Centauri. However, they are thrown into the titular dilemma by a stowaway.

Their ever-helpful servant is a robot, whose design was reminiscent of Robby's. With his spinning glass-dome head and light-up chest, the robot was best known for saying, "Danger, Will Robinson," to the young boy in the show, played by Billy Mumy. Actor Bob May portrayed the robot.

Lost in Space: The Motion Picture

In 1998, *Lost in Space* was remade as a theatrical feature, and an updated robot appeared. This mechanical man was designed and built by Jim Henson's Creature Shop, named after the late founder of the Muppets.

One of the supervisors for the bot's design and construction was Verner Gresty, who worked in England with twelve special effects magicians to make the three thousand-pound, eight-foot-tall robot.

Operated by a crew of four, the device could travel at fifteen miles an hour. According to Gresty, he studied the physiques of American football players to create the bot, which had huge shoulders, as if wearing shoulder pads.

One of the best things about the bot for the *Lost in Space* movie is that, unlike his predecessors, there is no one inside him making him move. In other words, this bot is a real bot.

Alien Botz

In 1964, the 20th Century Fox studios made a film called *The Earth Dies Screaming,* starring an army of alien robot creatures that invade hell-bent on the violent takeover of the world.

The movie, directed by Terrence Fisher, depicts nine-foot-tall robots that weigh almost a half ton apiece. They feed on electricity. They kill human beings with a mere touch.

Bullets won't stop them!

They also have the ability to reanimate the dead, thus frustrating a group of gravediggers and creating an army of living-dead zombies to fight for their nefarious cause.

Before the movie is over most of the human race, as is true of so many Cold War–era films, has been wiped out by the evil bots. The remaining handful of people are scrambling to both survive and fight back.

R2D2 and C3PO

The next cinematic robots to be clear-cut inductees into the Robot Hall of Fame made their first appearance in *Star Wars* in 1977. They were, of course, the lovable R2D2 and C3PO.

The movie, directed by George Lucas, won seven Oscars, and the robots still had no trouble stealing the show.

They were, like Robby, not machines as much as characters in the film, and their constant bickering, again like Robby, played them straight out of comic relief tradition.

The duo of R2D2 and C3PO also appeared in the next two entries in the Star Wars series, *The Empire Strikes Back* (1980), directed by Irvin Kershner, and *Return of the Jedi* (1983), directed by Richard Marquand.

All three were among the most successful motion pictures in Hollywood history.

Toon Bots

Two all-time classic robots came in animated features. There was the maid, Rosey the Robot, on the 1960s TV series *The Jetsons,* a show out of the Hanna-Barbera animations studio.

The fact that the Jetsons treated her like a member of the family—without ever forgetting, of course, that she was a domestic—offered a very optimistic view of robotics and their function in our lives in the future.

The other noteworthy robot character to come out of the world of animation was the metallic visitor from another world in the 1999 Warner Brothers animated film *The Iron Giant,* which is set during the time of the Cold War.

The story tells of a huge robot from outer space who has the capability to imitate anything. Will he imitate the kind mannerisms of the kind-hearted boy who befriends him or the mass-destructive power of the army's nuclear arsenal?

RoboCop

In the late 1980s, a new superstar bot appeared on the silver screen. *RoboCop* was about a Detroit police officer from the future who is brought back from near death with his own brain and a robot body.

The robot for this film was created by special effects wizard Rob Bottin. The film was directed by the Netherlands' biggest director, Paul Verhoeven, and released by Orion Films in 1987.

RoboCop's violent form of dishing out justice made this a super smash at the box office, big enough to spawn two sequels, *RoboCop II* (1990), this time with Irvin Kershner directing, and *RoboCop III* (1991) with Fred Dekker directing.

In 1994, *RoboCop* became a short-lived syndicated television series.

Number Five

A lot of bot-philes will tell you that their all-time favorite movie robot is the lovable Number Five from the 1986 movie starring Ally Sheedy and Steve Guttenberg, *Short Circuit.* Number Five has a great personality and that has a lot to do with why he is such a favorite, but his appearance has a lot to do with it, too. His tubes and wires form his human-like body parts. Number Five, like all good movie robots, has the capability to be dangerous as well. He is a battlefield weapon that gets hit by lightning and, because of the titular result, he thinks he is a person with a mind of his own. During the rest of the movie, brilliant young scientist Steve Guttenberg tries to get Number Five back to normal before mass destruction ensues.

Robot Man

No one today is doing more to raise consciousness and preserve the memory of film robots than Fred "Robot Man" Barton. Having fallen in love with Robby the Robot at age four when seeing the movie *Forbidden Planet,* the forty-three-year-old Barton went on to be the head of a Hollywood-based company known as Fred Barton Productions, Inc. The company is the world's number-one producer of museum-quality movie-and-TV robots.

Barton's works are highly sought after by robot collectors. Other robots he has built exact facsimiles of include Robot Model B9 from the TV show *Lost in Space,* Gort from the movie *The Day the Earth Stood Still,* C3PO and R2D2 from *Star Wars,* and Maria from the classic silent film *Metropolis.*

Today Fred is known as Robot Man. Fred built his "Robby" in the early 1970s, and it was so precisely matched to the original that he was called upon to do restoration work on the original robot that had been in the movie.

Barton's Robby lived in a museum at the time, but after his restoration, which made him as good as new, he was sold to a private collector.

Hollywood and Real: One and the Same

Today, the sets of big-budget Hollywood movies is a place to find real robots. Movies like *Mighty Joe Young* and the *Jurassic Park* films use robotics to portray actual oversized animals.

The other great place to go see real robots is at Disney World in Orlando, Florida, where robotics are frequently used as part of the exhibits and rides. In the Disney-MGM Studios, robots are an intricate part of The Great Movie Ride (where, for example, a robot Humphrey Bogart delivers his "hill of beans" speech to a robot Ingrid Bergman), and Star Tours, which simulates a trip through space.

At the Magic Kingdom, the Hall of Presidents contains a robot facsimile of each of the presidents. They all appear onstage, blinking and moving at once. The "Extra-terror-estial Alien Encounter" and "Buzz Lightyear's Space Ranger Spin" rides also include impressive use of robotics.

Two Visions

We find that there are two distinct visions for robots of the future in our Hollywood history.

There are those, dating back to the time of the Great Depression when times were tough, that robots are here to do jobs that were normally handled by people and are therefore evil.

Then there are those who see robots as lovable and subservient to the human race, companions to be loved and cherished, useful machines that will make all of our lives easier in the future.

So which will it be? It's safe to say that both sides, to an extent, will see their visions come true.

3

A History of Real Robots

The word "robot" is derived from the Czech word *robota,* which means slave and dates back to the thirtieth century B.C. Thoughts of what we would today consider robots date back as far as the ancient Egyptians and Greeks, both of whom wrote of statues that could talk and move.

Early Conceptions

Hero of Alexandria, circa 100 B.C., suggested that one day machines would do simple tasks for man, such as open the door or pull a plow.

It has been said that Leonardo da Vinci, the same man who painted the *Mona Lisa,* also built a mechanical lion that could walk and roar.

Robots, even back in the early nineteenth century, were thought of as monsters, as examples of man tampering in God's domain, feelings expressed with particular fleshy ghoulism in Mary Shelley's classic gothic horror novel of 1817, *Frankenstein.*

But it is at the beginning of the twentith century that robots are dealt with in the imagination much in the same form as we envision them today. As we learned in the previous chapter, "A History of Hollywood Robots," the word *robota* was first used to mean a mechanical man in the works of Czech playwright Karel Capek, a doctor of philosophy who wrote about a downfall of society that comes after all labor is done by soulless mechanical creatures.

Asimov's Laws of Robotics

The noted science-fiction writer Isaac Asimov had robots as part of mankind's future down to a science. His uses of mechanial creatures in his work, which were numerous, all followed the same rules.

He wrote about those rules in a piece called "Laws of Robotics." Asimov's rules were: 1.) A robot is not allowed to hurt a person or allow a person to be hurt; 2.) a robot must obey people unless it is being ordered to hurt another person; and, 3.) a robot must practice self-defense unless protecting itself will hurt a human being.

Asimov used robots in *I, Robot* (1950), the trilogy *The Caves of Steel* (1954), *The Naked Sun* (1957), and *The Robots of Dawn* (1983), as well as other stories.

The truth was, outside of Hollywood, there wasn't a lot of consideration put to building robots for the cause of evil. Robots were built to serve man, and Asimov's point was that no matter how sophisticated our technology might become, that was always going to be the case.

Reality

The first real robots, mechanisms to perform tasks that normally had to be done by people, were simple machines built in the mid- to late-nineteenth century.

Swiss artisans Jacques de Vaucanson, Pierre and Henri-Louis Jacquet-Droz, and Henri Maillerdet were pioneers when it came to inventing these types of machines. By the end of the century, this foursome had invented automatons that could play musical instruments, sign a signature that looked just like yours, or draw a picture of your house.

This really was the birth of applied robotics, even if there weren't any earthshaking changes in society being brought about by the actual inventions. In 1862, Alphonse Beau de Rochas created the most useful robot, which performed mundane and dangerous tasks associated with the operation of the first combustible engines.

Elektro and Sparko

In 1939, two real robots appeared on the scene that forever had an impact on how we think robots should look. The place was the New York City World's Fair and the robots lived in the World of the Future exhibit.

They were Elektro, who could walk both forward and backward and had a vocabulary of seventy-seven words. The other robot was Elektro's dog, Sparko, which could also walk and bark and wag his tail.

1946

As soon as the second world war ended, the great brains of America, with no war to win, went to work on modernizing the world. Within a year of peacetime, George Devol had a patent for a general-purpose playback device for controlling machines. The device used a magnetic process recorder.

That same year, the computer had its big public coming-out party. American scientists J. Presper Eckert and John Mauchly constructed the first large electronic computer. It was called the Eniac and it lived at the University of Pennsylvania. A second computer, the first general-purpose digital computer, dubbed Whirlwind, solved its first problem at M.I.T.

Telemetrics

Robotics as an aid to man sure took a giant leap during the years following World War II. Along with the development of computers came the development of a science known as telemetrics.

First developed by Argonne National Laboratories, telemetrics involved the creation of mechanical arms—as similar as possible to a human arm—that could lift, turn, and replace objects.

In 1961 General Motors began to use such mechanical arms to assemble its automobiles along the assembly lines. Different arms were used to paint, weld, and lift heavy parts.

If any machine that saves man work is a robot then every personal computer is a robot, often saving its owner steps and hours every day. We want our robots to walk and talk like people—although I suppose we would be satisfied if our bots just rolled along on wheels like Rosey on *The Jetsons*.

Modern Era

About thirty years ago a robot named Shakey, produced by SRI International, became the first mobile robot controlled by artificial intelligence. In 1974, an invention called The Silver Arm became the first robotic appendage capable of small-parts assembly.

Five years later, a robot known as the Standford Cart, using entirely preprogrammed commands, successfully crossed a chair-filled room without being assisted by a human being. It did this with a TV camera attached to a computer.

The camera did the seeing and the computer did the reacting.

The process of replacing human laborers with machines was called automation. Today's robot research continues to push toward the best ways for machines to serve man. But now the technology of computers, lasers, and radar have been added to the mechanical and electrical technologies of forty years ago.

HOAP-1

On September 12, 2001—in an announcement all but ignored because of the tragic events of the previous day—Fujitsu Laboratories debuted a humanoid robot designed to function as a two-legged domestic servant.

The prototype, called HOAP-1, will serve as a test bed for advanced research in robotic technologies. Fujitsu hopes to sell as many as one hundred of the robots at $41,400 apiece. Those sales aren't expected to go to rich people who want a mechanical butler to answer the door for them—although that could be arranged—but rather to schools that are involved in robotic research.

"It is meant for universities who want to do research on two-legged walking and robotic movement," said Scott Ikeda, a spokesperson for Fujitsu. "They can use it for research, such as developing algorithms for more natural two-legged movement."

The robot has flat feet, a wide chest, and a visorlike head. It stands nineteen inches tall, and weighs about sixteen pounds—so it can easily be moved from place to place when it is turned off. The robot is controlled from a personal computer.

Robomower

In Israel, a company called Friendly Robotics is marketing a bot called Robomower, which cuts your lawn while you sit on the porch sipping a tall, cool one. It runs on batteries, but it does have its problems.

It is not programmed to efficiently go back and forth or in diminishing concentric circles to guarantee complete lawn coverage, but rather travels randomly over the lawn until all of the grass has been cut.

It sometimes takes longer than it should, as the mower goes over areas where the grass has already been cut, but it sure beats standing behind the thing, pushing it. The mower won't wander off your lawn, either.

To keep the bot from going into the street or heading down the sidewalk to the corner tavern as its owner might, wires are placed at the edges of the lawn. The wires transmit signals that the Robomower senses.

So far Friendly Robotics have sold fifteen thousand Robomowers at $800 apiece.

Home Surveillance

A company called Probotics in Pittsburgh, Pennsylvania, has a robot vacuum cleaner on the market, which works along the same lines as the Robomower. Probotics also offers a "Spy-Cye" robot that keeps an eye on your house while you are away.

The Spy-Cye has a video camera mounted on it and roams around your house looking for intruders. You can monitor what the robot is "seeing" on your PC from wherever you are. A similar home surveillance device is expected soon from the Somerville, Massachusetts, company iRobot.

Since both products have difficulty with stairs—and sometimes even with the edge of carpet—it is not recommended that you use the bot as a substitute for a household alarm system. The technological lapse that must be overcome before security guards around the world will have to start to worry is, as of now, that the bots are not equipped to comprehend three dimensions. According to Illah Nourbakhsh, assistant professor of robotics at Carnegie Mellon University in Pittsburgh, "No [mass-produced] robot can follow you around stairs or curbs. It's hard getting around without legs in the human world."

According to Joe Engelberger, a pioneer in the U.S. robotics industry, single-function robots will never be appealing to a mass market, because those sorts of robots are not used frequently enough to make them worthwhile. Engelberger says mass-market bots must be capable of doing many household chores so that they are useful around the clock.

Everyone agrees that we are between ten and twenty-five years away from such a product being available to the public.

SDR-3X

Sony's prototype humanoid robot, which stands about twenty inches tall, is called the SDR-3X. The SDR stands for Sony Dream Robot. It has twenty-four joints in its body and appendages—so it can walk, kick, and dance.

Sony's robot is noteworthy because it contains a small-sized actuator that combines a motor, gearbox, and associated circuitry, all combined into a small package. The SDR-3X has three types of actuators that differ by the torque, or turning power, each one can provide.

According to Toshitada Doi, a corporate executive vice president at Sony and head of the company's Digital Creatures Laboratory, the SDR-3X has what is called synchronized body movement.

In other words, the SDR-3X can move the upper half of its body to counteract the movement generated by the lower half and so maintain balance. Dual RISC (reduced instruction set computer) processors continuously crunch data from sensors mounted throughout the robot to ensure it stays balanced. They are powerful enough to enable a walking speed of fifteen meters per minute.

The bot also features voice and image recognition. It "sees" with the help of a color 180,000-pixel resolution charge-coupled device pickup similar to those currently found in digital still cameras.

If you are eager to purchase one, cool your jets. The prototype is not going to be mass-produced for consumers any time soon. The bot might be used as a technological base for future commercial projects.

Sony has made it clear that they see robots as a source of entertainment rather than usefulness in the home. They proved that in 1999 when they sold 150,000 copies of their "Aibo" robot dog.

Creative Robotics

Over in England, state-of-the-art robot research is being done at Creative Robotics at Magna. The team of roboteers is led by Professor Noel Sharkey of the Computer Science Department at Sheffield University.

By Easter 2002, Sharkey and his crew had exhibited a colony of self-sufficient and evolving robots known as the Predator and Prey robots. Fifteen bots survived by feeding on light sources.

They will be hunted by five Predator robots, which will maintain their energy by catching and sucking "life" from the Prey. The successful survivors of both Predator and Prey will go on to breed and so the colony will evolve.

During the summer of 2002, Creative Robotics is expected to debut their "flyborgs," which are independent, flying bots that behave like birds and insects, flocking and swarming, taking their lead from the robot in front.

New Controller

Cutting-edge robotic research and the world of robot combat are connected. Acroname Inc.'s BrainStem is the newest in the line of microcontrollers. Aside from being a great microcontroller for autonomous robots and other industrial applications, it is a great controller for fighting bots as well.

Says Acroname's Jim Sincock: "BrainStem is an iPaq with a TCP/IP wireless link and a joystick for teleoperation. There are several benefits to this approach. One major benefit is that the relay doesn't care what operating system or even programming environment is connecting to it. This means that a Web page, Java program, or strange operating system can easily connect to and control the BrainStem from anywhere on the Internet (provided the networks are set up properly).

"Another benefit is that programs, individual commands, and even debugging can take place remotely, potentially while your robot is moving about. We feel that this method of control will allow for more refined control of the robot, and there is the ability to begin to incorporate sensors and forced feedback. As more builders begin to incorporate sensory systems to their bots, perhaps a fully autonomous class will enter the world of robotic combat."

To the Edges of the Galaxy

Today robotic space exploration has allowed us to see to the edges of the galaxy and beyond. We've wandered the surface of Mars and flown by Pluto. We've seen the rings of Saturn and Jupiter's red eye close up.

And we've invented a fun new sport. Robot against robot, in the pit, fighting to the death. It somehow combines our finest and most refined qualities with our basest animalistic instincts.

Cool.

4

Other Contests

We learned in the "History of Robotic Combat" chapter that there are three battling bot TV shows on in the United States: *BattleBots, Robot Wars: Extreme Warriors,* and *Robotica.* But that is definitely not the extent of robotic competitions.

First, there is robot soccer. The fifth annual **RoboCup** tournament was held in Seattle, Washington, the first time the event was held in the United States. One hundred eleven teams from twenty-three countries entered the event, in which the players are robots, the field is a Ping-Pong table, and the ball is a golf ball.

RoboCup teams compete in four leagues, depending on the type of robot: small, about the size of a softball; middle, about twenty inches in diameter; four-legged Sony Aibo robots, about a foot high; and the computer-simulation category, which eliminates the hardware from the equation.

The robots are not operated by radio control but are preprogrammed to react in certain ways to specific situations. But, in the chaos of a soccer game, the robots inevitably run into situations that they haven't been programmed for.

"We can't preprogram everything," said Manuela M. Veloso, associate professor of computer science at Carnegie Mellon University and chairwoman of RoboCup 2001. "We make them capable of handling a wide spectrum of situations. When we see the game, we get surprised by what they do."

Relevant objects, like the goals, players, ball, and walls of the field, are different colors so that the robots can distinguish them from one another. The players can reach speeds up to five feet per second.

The small robots have a video camera above the field that keeps track of the ball, goals, and all the robots. That information is fed to an off-court computer, which guides strategy and tells the robots how to move.

In the middle and four-legged leagues, the robots carry their own cameras, allowing each to operate autonomously. As technology improves, so does the quality of the game. Although the competition is meant for fun, it is hoped that the technologies being developed for RoboCup will have applications that can help mankind.

Here are some examples of other competitions that pit bot versus bot, although not always in combat. Some competitions measure just how useful a bot can be. For example, at the 10th Annual **AAAI Mobile Robot** competition held during August 2001 at the Washington State Convention and Trade Center in Seattle, Washington, robots—like those you met in the Preface to this book—competed to locate "injured" humans in a theoretical disaster area.

For those of you whose favorite all-time bot is Rosey from *The Jetsons*, this competition also gauges bots' ability to function as a domestic helper, that is, to serve and interact with humans.

Each year, Acroname Inc. hosts the **Acroname Robotics Competitions and Expo** in Boulder, Colorado. The expo includes several autonomous robotic competitions as well as lectures and presentations from NASA and others working in the field of advanced robotics. It is a great opportunity for people to check out and participate in the field of autonomous robotics. The competitions that are held each year include:

The Acroname Obstacle Course: This competition was designed by Acroname and is exclusive to this expo. The course is designed for autonomous robots (no radio control), which must be programmed to use sensors to navigate the course. According to Acroname's Jim Sincock: "The robot must be able to wall follow and line follow as well as know how to continue on when the line or wall disappears. The course curves and climbs and includes sections such as: the 'Swiss Cheese section' where the robot's wheels could drop into a hole and get stuck; the 'astro turf hill,' which will test not only the power of the robot, but also the balance; and then there is the infamous ball pit. The pit drops off at a sharp angle and is filled with plastic golf balls. This is the final piece in the course and so far no one has built a robot that can make it all the way through it. Each year, a new piece is added to the course just to spice it up for the return contestants. The course is part of our LEGO Workshop, but is also open to other hand-built autonomous robots." More info, rules, and track specs can be found at http://www.robo02.com/obstacledetails.html.

LEGO Workshop: Check out http://www.robo02.com/workshopdetails.html for details on this competition.

Trinity Fire Fighting Pre-Run: This contest will be an exact replica of the huge contest held every spring at Trinity College in Hartford, Connecticut. This contest involves an autonomous robot navigating a small maze, trying to find a lit candle, putting the candle out, and returning to the "home base." This contest counts as a regional qualifier for the Trinity Contest, which is held each year in mid to late April. Details on the Trinity Contest can be found at http://www.trincoll.edu/events/robot/rules01.htm.

Line Following: This contest will be restricted to self-controlled robots created by enthusiasts sixteen years old and younger. In the past only LEGO-based robots were admitted. Now, any autonomous robot is eligible to compete. Says Sincock: "In other words, once turned on, the robot must be computer controlled without any human intervention from manual devices, such as remote controls." The contest involves a small ten-foot-by-six-foot arena with a one-inch-wide black line on a white background. Robots placed on the starting line need to follow the line as it wiggles around the arena. Robots are scored based on how far they can follow the line and the time it takes for them to complete the course. There are some tricky turns and obstacles along the way. For more details check out: http://www.robo02.com/linedetails.html.

Sumo: Robot Sumo is an extremely popular and fiercely competitive event in Japan, and is becoming quite popular in the United States. The robot sumo competition is a match of speed, skill, and raw power between two robots in a five-foot-diameter elevated ring. The objective is to locate the other robot without falling out of the ring, and once located push your opponent out of the ring. The first robot to touch the floor loses. The basic rules follow the All-Japan Robot Sumo Tournament. As with the other Acroname competitions, these robots must be totally autonomous. Says Sincock: "Once started they can not be controlled by the operator by any means." There are generally two weight classes, one kilogram and three kilogram. The robot must fit into a twenty-centimeter-square box, but there is no height restriction. The actual bout lasts a maximum of three minutes. If, after three minutes, robots are still in the ring, the bout is stopped and the "ref" decides on either a winner or a rematch.

In the **All Japan MicroMouse Contest,** which has been held annually at various locations in Japan for more than twenty years, robots race through a maze.

An interesting variation on the maze race comes at the **AMD Jerry Sanders Creative Design Contest,** which is held annually at the University of Illinois at Urbana-Champaign. In this event, autonomous or radio-controlled bots must navigate a maze while locating and collecting balls along the way and transporting them to a home area. The balls come in two sizes: light bocce balls and fifteen-pound bowling balls. The balls are placed both on the ground and on elevated surfaces to better test the bots' manual dexterity.

For those who might be interested in entering a bot in this maze contest, be aware that there is no line-of-sight visibility between the robots and their controllers while the bots are in the maze, so those bots that are not autonomous and rely on radio control will need video capabilities.

If space robotics are your thing, then you'll want to check out the **ASCE Lunar Robotics Construction Contest,** which is held biannually in Albuquerque, New Mexico. In this competition, robots must survive a simulated lunar landing, locate a simulated subsurface lunar tube, drill an opening into the tube, and mount an airlock in such a manner that it is airtight. The 2002 version of this contest will be held in conjunction with the Space and Robotics 2002 Conference and Exposition.

At the annual **Atlanta Robot Rally,** there are three contests:

1. An Open Contest, in which the builder chooses his own task for his bot to complete;
2. the Kit Robot Contest, which is also a self-chosen task, but for a kit robot; and
3. the Vacuuming Contest, in which bots are timed performing this household chore.

The show is sponsored by the Atlanta Hobby Robot Club.

For those of you who are Down Under and half-way around the world from most robot contests, check out the **Auckland Robotic Games** held at Tamaki Campus, University of Auckland, Auckland, New Zealand. There is a robot race, a treasure hunt, and contests for artistically designed robots and scientific innovation.

At the **AUVS International Aerial Robotics Competition,** held at Webster Field in St. Inigoes, Maryland, there is a Hostage Rescue contest, a

Nuclear Disaster contest (what fun!), in which bots must transport sensors to a radiation-contaminated area, and a Biological Emergency contest, in which bots return data from a quarantine area.

The **AUVS International Ground Robotics Competition** is held at various locations, with the next one, in July 2002, to be held in Orlando, Florida. In this show's main contest, autonomous ground vehicles navigate an outdoor obstacle course within a prescribed time while staying within a five-mile-per-hour speed limit.

Robots here are autonomous only. They may be between three and nine feet in length and between five and six feet tall, excluding antenna. Propulsion must be by direct ground contact and power must be supplied by combustible fuel or electricity. The best news for bot builders is that there is $20,000 in prize money. The bad news is that the competition is open to students only.

The **AUVS International Undersea Robotics Competition** was last held in July 2001 at the United States Naval Academy in Annapolis, Maryland. The next event is scheduled for July 2002, and will be held at Canada Olympic Park in Calgary.

This competition is for autonomous-only bots, which can not be greater than six feet long by three feet wide by two feet deep. They may not weigh more than 100 kilograms.

The contest is open to students only. Robotics clubs and organizations are excluded from participating unless at least 75 percent of their team are students. No outside hardware of any kind is allowed in this competition and that includes GPS satellites and tethered computers.

At **BotBash,** last held at the Confederate Air Force Base in Phoenix, Arizona, both autonomous and radio-controlled bots compete in battle, capture the flag, and obstacle-course contests.

Bots and builders are divided in expert and novice classes, and bots are divided into three weight classes: twenty-four pounds, fifty-five pounds, and eighty-seven pounds. The 2002 event will take place September 13–15 at the Army National Guard Armory on the campus of Arizona State University in Tempe.

The **Canada First Robotic Games** was last held during March 2002 in Toronto, Ontario, Canada. This competition is for remote-control bots only. Robots compete in the RoboBiathlon—that is, navigate an obstacle course and fire projectiles with accuracy.

The **Central Jersey Robot Conflict** is held annually at the ShoreCon

Convention Center in the Four Points Hilton in Cherry Hill, New Jersey. Featuring fights to the finish, the "Conflict" is for small radio-controlled bots only, ranging in weight from eleven pounds to twenty-four pounds.

The **CIRC Autonomous Sumo Robot Competition** for sumo wrestling autonomous bots will next be held at the Lakeview Museum in Peoria, Illinois. Bots here are divided into three weight classes: 3 kilograms, 1.5 kilograms, and 500 grams.

Of all the bot competitions in the world, there are few with names we like better than **Critter Crunch.** Both radio-controlled and autonomous bots are eligible for bot combat at this event.

Critter Crunch bots must be no longer than one foot cubed and the top weight limit is twenty pounds. This contest was created by the Denver Area Mad Scientists, a group that still governs the contests.

This is the original robot battle competition and predates Robot Wars and BattleBots by more than ten years. The competition is usually held in conjunction with the MileHiCon convention each year. Additional contests are sometimes held in other areas using the same rules.

DragonCon Robot Battles are held annually at the Hyatt Regency Atlanta and Atlanta Marriott Marquis in Atlanta, Georgia. The event is held as part of a science-fiction convention called DragonCon. Both autonomous and radio-controlled bots may compete in three weight classes, ten-, twenty-five-, and fifty-pound maximums.

The Dallas Personal Robotics Group has held the **RoboRama Competition** biannually since 1997. The event, held in Dallas, Texas, most recently at the Science Place, is open to bots under three hundred pounds.

Three events are held at RoboRama: the Quick Trip, for beginners, in which the bots move from point A to point B and back again; Line Following, in which bots must follow a line on the floor; and Can Can, in which bots locate and retrieve empty soda cans.

So, there are all sorts of options for fans and builders. If you can't make it to a taping of *BattleBots,* check out another robot competition, maybe a little closer to you. You're bound to have a blast.

5

BOTZ A to Z: An Encyclopedia of Battling Robots

A

ABADDON

Abaddon was eliminated in the opening round of the *Robot Wars* 3 (UK) tournament by Technophobic.

ABATTOIR

A superheavyweight veteran of *BattleBots,* Abattoir's primary weapons are counter-rotating wheel blades. Ray Scully, from San Anselmo, California, built the bot with the help of Helen Smith, Susan Schalit, and Matthew Johnston.

About Abattoir, Scully says, "Scary ideas with substandard construction. At least four welds held together with duct tape. Electrical systems melting as we speak. Control systems circa 1979, not Y2K compliant."

As far as influences are concerned, Scully says, "Einstein, Peter Garrett, Wile E. Coyote. Mark Pauline and SRL, The Seemen, the U.S. military-industrial complex, Mescaline. Not in that order. I get my fighting-dirty skills from my family. I figured out that everything is tougher than you expect outside of school. I found out that you can depend on your friends for all of the unimportant things and to eat all your food. And I gained my sense of patience and the ridiculous from corporate America."

A BOT NAMED SUE

I hear the train a-comin'. It's rollin' round the bend. Do you get the impression that Christopher Powell of Denver, Colorado, might be a Johnny Cash fan? His bot is named A Bot Named Sue and his team is known as The Team in Black.

Unfortunately, A Bot Named Sue fell into a burning ring of fire in the opening round of *BattleBots*' Season 3 lightweight tournament, losing a 38–7 decision to Carnage Raptor.

THE ADMINISTRATOR

A heavyweight built by Austin Carpenter of Edge Robotics in Germantown, Tennessee, The Administrator lost in the opening round of *BattleBots*' Season 3 heavyweight tournament to Greenspan, coming up on the short end of a 27–18 decision.

ÆON 00

Æon 00 was built by Thomas Ragaini and Team Big Fire of Kirkland, Washington. It was KO'd in the opening round of *BattleBots*' Season 3 lightweight tournament by Silver Bullet.

AFRICANIZED KILLER BOT

In Season 3 of *BattleBots,* this superheavyweight bot, built by Patrick Campbell of Stevenson Ranch, California, won a 27–18 decision over Cannibot in the first round but lost by a KO to Junkyard Cog in Round 2.

AFTERMATH

Aftermath was knocked out in the opening round of *BattleBots*' Season 3 middleweight tournament by Booby Trap. It is the creation of Steven West and Team Black Op West of Pleasanton, California.

AFTERTHOUGHT 2.0

Built in about a month by Jim Sellers and Steve Winter of Oregon, Wisconsin, Afterthought 2.0 is a lightweight veteran of the BattleBox. Battery powered, the bot's primary weapon is a disc that spins at 2,500 revolutions per minute. Depending on the opponent, Afterthought 2.0 can quickly switch discs to attack with a sharp-bladed disc or perhaps a disc with steel balls on chains attached to it.

AGAMEMNON

A product of Dan Danknick and Team Delta, Agamemnon weighs 100 pounds and operates on surplus 0.33-horsepower PMDC motors, with the power being provided by two 12-volt, 4ah gel cells. It is controlled via a five-channel FM radio.

Its primary weapons are gas-powered saws and an air punch. The latter weapon is operated using a twelve-ounce carbon-dioxide tank. Agamemnon's frame is welded angle steel. The armor is made of aluminum plate. It took nine months to build and cost $2,411.

Agamemnon has a bottle of 1,000-pounds-per-square-inch compressed nitrogen aboard. For competition that is regulated down to 150 pounds per square inch and switched through two high-flow air valves to control the action of the pneumatic punch.

The punch design originally started out as a spring-loaded single-shot harpoon. To the end of the weapon was added a titanium tip. It started life controlled with off-the-shelf R/C parts, but this was eventually interfaced with an advanced C&H flight stick to the transmitter. A Tekin R/C speed controller is used to switch the traction motors.

The bot also features a video feed thanks to an onboard micro-sized CCD camera that relays a signal through a video transmitter. The image is then shown on the insides of a pair of virtual-reality glasses so that the driver always has a robot's-eye view of the action.

Agamemnon made its debut at the 1996 Robot Wars competition, where it dominated the middleweight division.

The bot's name is derived from the television show *Babylon 5*. On that show the *Agamemnon* is an Omega-class destroyer. See also www.teamdelta.com.

AGENT ORANGE

Agent Orange weighs 174 pounds and was built by England's Peter Duncanson and Team Spin Doctor. It runs on wheels and operates on electricity. Its multiple weapons include a hammer/ax, a spike, and a pincer. The weapons operate on both electric and hydraulic power.

In the *Robot Wars* 3 (UK) tournament, Agent Orange defeated Max

Damage in the opening round of its heat, then lost in Round 2 to Blade.

AGGRAVATED ASSAULT

Rob Gibson-Dugger and Real Good Robotics of New York City constructed Aggravated Assault, which KO'd Nicebot in the opening round of *BattleBots'* Season 3 lightweight tournament, then was KO'd itself by Malice in Round 2.

AGGROBOT

Aggrobot defeated Daisy in the opening round of its heat at the *Robot Wars* 3 (UK) tournament. In Round 2 it defeated Razer, then was itself eliminated from the competition by Blade in the heat finals.

AGRIPPA

Agrippa, which took about seven months and $2,411 to build, weighs 207 pounds and is driven by a surplus 1.3-horsepower DC traction motor. A product of Dan Danknick and Team Delta, it is powered by fourteen six-cell nicad packs and controlled via a four-channel PCM radio.

Its main weapon is a linear blade array. The weapon is powered by an eight-ounce carbon-dioxide tank and a Hawker battery. The frame was constructed of one-inch welded square tube steel. The armor is hybrid carbon fiber over aluminum/magnesium. See also www.teamdelta.com.

AGSMA

Not a common lung ailment, but a fighting robot! Agsma, which was built by John Neilson of El Cajon, California, was defeated in the opening round of *BattleBots'* Season 3 middleweight tournament by Die Fledermaus by a score of 26–19.

ALABAMA SLAMMER

Riding a crimson tide comes Alabama Slammer, the creation of Richard Nickel of Huntsville, Alabama. In *BattleBots'* Season 3 middleweight tournament, Alabama Slammer knocked out Bot Mulcher in the opening round, defeated Enumscratch by a score of 37–8 in Round 2, outpointed

Evil Con Carne by a score of 37–8 in the third round, then lost by a KO to Sunshine Lollibot.

ALEXANDER

Another bot from Dan Danknick and Team Delta. Eight months under construction, Alexander cost $2,203 to build. Its armor is comprised of aluminum diamond plate, while its frame is made of one-half-inch welded square tube steel. Its weapons are a milling cutter and an air punch, which are powered by nicad batteries and an eight-ounce carbon-dioxide tank.

It is controlled via two four-channel PCM radios, powered by two 12-volt, 7ah gel cells, and driven by two 0.5-horsepower Sullivan motors. It weighs 102 pounds. See also www.teamdelta.com.

ALEXANDER II

Weighing 100 kilograms, Alexander II is driven by two Bosch GPA750 motors. Its power is derived from two 12-volt, 13ah Hawker batteries. Controlled by one five-channel PCM radio, its weapon is a blade powered by a 5-horsepower milspec motor drive. The frame is constructed of one-half-inch welded square tube steel and its armor is made of titanium plate.

ALIEN II

This lightweight veteran of the *BattleBots* competition is battery powered and built on wheels. Daniel Rupert, Greg Sharp, and Edward Liu of San Diego, California, built Alien II over a three-month period.

ALIEN GLADIATOR

A middleweight bot (115 pounds) built by Daniel Rupert and Team Malicious. Alien Gladiator lost to Deadblow in the *BattleBots*' Season 1 middleweight tournament, despite the fact that Deadblow's pickax weapon snapped off in the middle of the contest due to metal fatigue.

ALL TORQUE

At the *Robot Wars* 3 (UK) competition, All Torque defeated Crippler in the opening round of its heat, then was eliminated in the second round by the eventual heat winner, Thing II.

ALL WEATHER TIRE BOT

A participant in BotBash, All Weather Tire Bot is made up of snow chains on spinning tires and weighs slightly more than 100 pounds. It took four months to build and cost about $1,000. Both its motor and weapon are battery powered. The builders are Ryan Porchia and Travis Wood.

ALLY GATOR

At the *Robot Wars* 3 (UK) competition, Ally Gator chomped its way to victory over Corporal Punishment in the first round of its heat, then was eliminated from the tournament in the second round by Beserk II.

ALPHA RAPTOR

Alpha Raptor is a lightweight, tipping the scales at only 56 pounds. It has wheels, and both its drive and weapon are battery powered. The primary weapon is a lifter arm. The main builder is Chuck Pitzer from Phoenix, Arizona. The other members of Team Raptor are Wayne Wray, James Wilkinson, and Chris Harriman.

Chuck is a mechanical engineering student who has been "intensely trained" on Pro/E and Mechanica design and simulation software.

Says Pitzer: "I built my first robot in college due to a fascination with control systems and robotics. It was named Hyena, an autonomous metal-detecting robot—that is, a scavenger robot." His first exposure to the sport came from watching the original *Robot Wars.* He says that the sport is much more competitive now than when he first got started. "There's not quite as much comraderie as there was in the beginning. People are starting to really look at the return on investment and profitability now."

Chuck's words to live by are "Chance favors the prepared mind," a quote from Louis Pasteur.

At *BattleBots*' Season 1 lightweight tournament, Alpha Raptor defeated Serial Box Killer, future-champion Dr. Inferno Jr., and Tentoumushi before being sliced and diced in the finals by Backlash.

ANKLE BITER

It's pesky, it's annoying, it's Ankle Biter, the brainchild of John McKenzie and Team Odin of Ukiah, California. Ankle Biter was knocked out in the opening round of *BattleBots*' Season 1 middleweight tournament by Pressure Drop. In this match, Ankle Biter appeared to be getting the best of it when it became wedged under the spikes that line the arena and was unable to continue.

In *BattleBots*' Season 3 tournament, Ankle Biter was also eliminated in the first round, this time by F5.

THE ANNIHILATOR

Damian Quintana and the Bulldogs may sound like a good name for a rock-and-roll band, but it is actually the name of the creators of The Annihilator and his team, all from Fresno, California. The Annihilator lost a 32–13 decision in the first round of *BattleBots*' Season 3 middleweight tournament to Fast Forward.

ANORAKAPHOBIA

Anorakaphobia defeated Miss Ile in the opening round of its heat at the *Robot Wars* 3 (UK) tournament, then was itself eliminated from the competition in Round 2 by Big Cheese.

ANUBIS

Anubis is a superheavyweight with a carbide metal saw blade as its primary weapon. The bot also has four lifting arms and steel plating. Built by Glen Sears of Woburn, Massachusetts, with the aid of Adam Godbout and Aaron Puncotte. It took them a full year to design Anubis and two months to build.

In Season 3 of *BattleBots,* Anubis won a 24–21 decision over DooAll, then lost in the second round to Phere, by a score of 27–18.

ARCHANGEL

Antoine Trabulsi and Genesis Robotics of Kitchener, Ontario, Canada, fathered Archangel, which lost a 30–15 decision to Rhode Hog in the opening round of *BattleBots*' Season 3 lightweight tournament.

THE ARCHDUKE

John Howard and Team Poison Fists of Pittsburgh, Pennsylvania, created The Archduke, which performed well at *BattleBots*' Season 3 lightweight tournament. It won a 25–20 decision over Ramming Speed, then outpointed Chimera 32–13. A forfeit pushed it into the fourth round, where it came up on the short end of a 31–14 decision to Mecha Tentoumushi.

ARIES

John Crain of Slate Hill, New York, built Aries, which was KO'd in the opening round of *BattleBots*' Season 3 lightweight tournament by Trilobot.

ARMADILLO

Armadillo was turned into roadkill by Double Agent in the first round of *BattleBots*' Season 3 middleweight tournament. Built by Mike Kalkwarf and his team, Creative Anarchy, from Roseville, California.

ARMAGEDDON

A superheavyweight veteran of *BattleBots*, was built by Doug Groves of Painesville, Ohio. In Season 3 of *BattleBots*, Armageddon defeated G.O.R.T. in the opening round, then lost a squeaker of a decision to Half Gassed in the second round, losing 24–21.

ARMORGEDDON

Armorgeddon is a heavyweight, at 210 pounds, and its strategy is clear at first sight: It has wedges on all four sides. Armorgeddon has a vacuum impeller to increase traction by creating a low-pressure area under the machine.

The bot also has a nine-inch piston powered by a 2,000-pounds-per-square-inch nitrogen bottle capable of exerting over 1,000 pounds of force through an appendage that disturbingly resembles an eyeball.

When Armorgeddon's outer shell is removed there is two inches of ground clearance for obstacle courses.

It was built by thirty-three-year-old engineering department manager Jason Giddings and twenty-five-year-old Scott Vandehey, of Hillsboro, Oregon. *Robotica II* was their first bot competition. Another team member is Jodi Giddings.

ARMOUR GEDDON

At the *Robot Wars* 3 (UK) competition, Armour Geddon was eliminated in the opening round of its heat by Onslaught.

ATILLA THE DRUM

At the *Robot Wars* 4 (UK) tourney, Atilla the Drum lost in the opening round of its heat to Medusa 2000.

ATLAS

Atlas not only couldn't hold the world on its shoulders, it couldn't even get past the opening round of its heat at the *Robot Wars* 3 (UK) tournament, losing to Challenger II.

ATOMIC

At *Robot Wars* 4, Atomic defeated Indefatigable in the opening round of its heat, outpointed King B3 in the second round, but then lost in the heat finals to Chaos II.

ATOMIC WEDGIE

Superheavyweight Atomic Wedgie, with its wedge attack along with spinning tri-foils, can turn your boxers into thongs until you're mighty thore. Try to fight back and you smack into Atomic Wedgie's titanium armor, which is 4⅛ inches thick.

Both drive and weapons are battery powered. The bot's spinning weapons are nuclear tri-foils with tungsten tips. They spin at 1,500 revolutions per minute. The bot moves on tank treads both on the top and the bottom, so it can't be flipped over.

The bot was built by Robert Everhart, Charlie Payne, and Tom Corrie—with the help of Robert Everhart III, Bradley Bowan II, and Kevin Maze—all from Columbia, Maryland.

In *BattleBots*' Season 2 superheavyweight competition, Atomic Wedgie advanced all the way to the finals, proving itself to be among the best of the biggest. Atomic Wedgie started the tournament by eliminating the defending champion, Minion—flipping Minion onto its back in the process. In the second round, Toro was Atomic Wedgie's victim. It beat Revision Z in the semis and then lost to Diesector in the finals.

Atomic Wedgie lost in the opening round of Season 3 of *BattleBots,* being KO'd by Techno Destructo.

AVATAR

Built by Joseph Butler and Project X of Tulsa, Oklahoma, Avatar was knocked out in the opening round of *BattleBots*' Season 3 middleweight tournament by Li'l Chipper.

A.W.O.L.

A.W.O.L. was away without power after it was KO'd in the opening round of *BattleBots*' Season 3 lightweight tournament by I-Will-Smack-U. This runaway bot was built by Aaron Horrocks and Team Black Op Six of Pleasanton, California.

AXIOS

At the *Robot Wars* 3 (UK) tournament, Axios lost its first match to the eventual heat winner, Panic Attack, and went home.

B

BACCHUS

Built by a team from Lafayette College in Easton, Pennsylvania, Bacchus did very well in *BattleBots*' Season 3 heavyweight tournament, winning four big matches before being knocked out of the competition. Bacchus defeated PyRAMidroid by a 32–13 decision, KO'd License to Kill and Blendo, and won a 39–6 decision over Junkyard Offspring before losing a 25–20 decision to Voltronic in the fifth round.

BACKLASH

Jim Smentowski, Backlash's builder, lives in Novato, California, which is about forty miles north of San Francisco. About himself, Jim says, "I am married, have a dog, a cat, two birds, and lots of robots. I love computers, electronic gizmos, car audio, power tools, and blueberries. I've created a Web site (robotcombat.com) for robot combat enthusiasts and other creators, hopefully to be able to educate, inform, and generally show off all that I can about robotic combat."

Jim is a special effects wizard for the movies in real life. He works for Industrial Light & Magic, the people who did the special effects for *Star Wars.*

At *BattleBots'* Season 1 lightweight tournament, Backlash was the class of the field. On its way to winning the Giant Nut, it shredded foes, repeatedly sending pieces of bot flying across the BattleBox. It defeated Disposable Hero, The Crusher, Endotherm, Das Bot, and, in the finals, Alpha Raptor.

Backlash won second place at the *BattleBots'* Season 2 lightweight tournament in 2000. At that tournament it knocked out Rampage in the opening round (which did not air on television), won a decisive 43–2 victory over Scrub in its second matchup, knocked out Sallad in the quarterfinals, scored another KO over Toe Crusher in the semis, then was knocked out—in under ninety seconds—by Ziggo in the finals.

At the Season 3 *BattleBots* tournament, Backlash knocked out Matts Bammer in Round 1, then was KO'd by Skid Mark in the second round.

BACKSTABBER

Backstabber got its comeuppance when it was sliced by Razer and eliminated from the competition in the opening round of the *Robot Wars* 3 (UK) tournament.

BAD ATTITUDE

The leader of Team Attitude is Thomas Petruccelli of San Jose, California. Tom received assistance in Bad Attitude's construction from Robert Borck and Bonnie Petruccelli. Bad Attitude's strong point is its speed. It is a wedge-shaped bot that is battery powered and built on wheels.

In *BattleBots'* Season 1 middleweight tournament, Bad Attitude was eliminated in its opener by Blade Runner.

Bad Attitude KO'd Garm in the first round of *BattleBots'* Season 3 tournament, then lost a close 24–21 decision to T-Wrex in Round 2.

BAD BADGER

Badgers? Badgers? We don't need no stinkin' badgers—not, that is, unless you're talking about Bad Badger, a heavyweight bot built by Ryan Lufkin of Salt Lake City, Utah.

In *BattleBots'* Season 3 heavyweight tournament, Bad Badger defeated Psycho Splatter by KO, then lost a lopsided 42–3 decision to RoboSapien in the second round.

BAD BONE

Buh-buh-buh-buh-bad! Will Keels and Team Bone of San Diego, California, built Bad Bone, which was KO'd by Sisyphus in the opening round of *BattleBots*' Season 3 lightweight tournament.

BAD DOG

Bad Dog—built by Kenneth Benner of Lake Mathews, California, along with team K-NINE—knocked out SledgeJammer in *BattleBots*' Season 3 heavyweight tournament, then was KO'd itself in the second round by Towering Inferno.

BAD HABIT

The offspring of Ray Alderman and Team Widget of Sunnyvale, California, Bad Habit KO'd I-Will-Smack-U in the opening round of *BattleBots*' Season 3 lightweight tournament, followed that up with a 23–22 victory over Reaper in Round 2, then was KO'd in Round 3 by Dr. Inferno Jr.

It is a wedge-shaped bot that is battery powered and built on wheels. Its primary weapon is its own kinetic energy—that is, it's a rammer.

Bad Habit took approximately three months to build, and runs no matter which side is up. Ray has built a switch into his transmitter so that, with one flick, he can reverse his controls in case the bot is flipped over.

BALLBREAKER

A superheavyweight built by Shawn Ouderkirk of Colorado Springs, Colorado. Ballbreaker lost in the opening round of Season 3 of *BattleBots* when it was KO'd by Flailbot.

BAMBINO

Built in the shadow of Cape Canaveral and the Kennedy Space Center—in Titusville, Florida—by Steve Wilson, Bambino outpointed Mr. Snoogglepuss 28–17 in the opening round of *BattleBots*' Season 3 lightweight tournament, then was clobbered 39–6 in Round 2 by Sparky.

BANG

Onomatopoeia isn't my favorite when it comes to bot names, so I wasn't overly disappointed when Bang—built by Troy Alton of Edmonton,

Alberta, Canada—got whupped by Village Idiot in its opening match in *BattleBots'* Season 3 middleweight tournament.

BATRAY

Ted Shimoda and Team Ray of Pasadena, California, are responsible for this bot, which was eliminated by Kraken in the opening round of *BattleBots'* Season 3 middleweight tournament.

BATTLERAT

Team BattleRat is headed by Ted Walters of Point Pleasant Beach, New Jersey. In competition, BattleRat KO'd Massacre in the opening round of *BattleBots'* Season 3 heavyweight tournament, won a 33–12 decision over Crash Test Dummy in the second round, then lost a 34–11 decision to Tazbot in Round 3.

Walters appeared on NBC's *Today* show during the summer of 2001 and told host Matt Lauer that he had watched the fighting robot shows on TV and had decided that he wanted to build one of his own. Walters took tips from Christian Carlberg's Web site and BattleRat was the result—a combination wedge and flipper. The bot cost between $8,000 and $10,000.

BATTLESCAR

Built by Brad Broman of Linn, Missouri, this superheavyweight lost in the opening round of Season 3 of *BattleBots.* Battlescar was defeated in a 27–18 decision to Trimangle.

BEAST OF BODMIN

At the *Robot Wars* 3 (UK) tournament, Beast of Bodmin was the winner of its opening heat, scoring consecutive victories over Crusher, Onslaught, and Invertibrat. In the semifinal heat, Beast of Bodmin defeated Blade in the opening round, but then lost in its attempt to make it into the tournament's final four to Steg-a-Saw-Us.

B.E.H.EMOTH

Louisville, Kentucky's, Eric Hilgeford built B.E.H.emoth, which knocked out Neb in the first round of *BattleBots'* Season 3 middleweight tournament before losing in Round 2 to Eraser 28–17.

In the *Robot Wars* 3 (UK) tournament, it defeated Shark Attack in the opening round of its heat, KO'd General Carnage in the second round, and then was eliminated by Pit Bull in the heat finals.

BENDER

Bender is the brainchild of Burbank, California's, Eric Stoliker, along with team members Brandon Nichols, Dan Wang, and Jim Snook. This heavyweight bot took five months to build.

A two-inch spinning blade is its primary weapon. Among its special features is an adjustable rear caster height. In *BattleBots*' Season 3 heavyweight tournament, Bender lost a 35–10 opening-round decision to Omega-13.

BERSERKER 2000

About Berserker 2000, main builder Joey Sala of Norman, Oklahoma, says, "Our spinning tire from above is mounted to the base of our robot using an electric clutch. The tire is used like a flywheel to store energy, which can be released by locking the clutch and transferring the angular momentum to the base, which has weapons."

That armored tire, the bot's primary weapon, is thirty-six inches in diameter and spins at 600 revolutions per minute. The bot, which took eight months to build, also has interchangeable weapons, so that it can adapt depending on its opponent.

BESERK II

At the *Robot Wars* 3 (UK) competition, Beserk II never did learn how to spell berserk, but it did defeat Tut's Revenge and Ally Gator before being eliminated in the finals of its heat by Hypnodisc.

At the *Robot Wars* 4 (UK) tourney, it was defeated in the opening round of its heat by the eventual heat winner, Tornado.

BETA RAPTOR

Beta Raptor is the lightweight entry in *BattleBots* tournaments from Chuck Pitzer, Jon Kiddlesrud, Wayne Wraye, Chris Harriman, and James Wilkinson of Chandler, Arizona. Beta Raptor took two months to design and two weeks to build.

It is battery powered—operating on twin cobalt 90 motors, which provide 1 horsepower apiece—and runs on wheels. Its primary weapon is a lifter arm, a linear-actuator driving arm capable of pressing one ton. It also features spring-steel-reinforced body armor with titanium armor on the sides.

BIG BROTHER

At the *Robot Wars* 3 (UK) tournament, Big Brother was the winner of its heat, scoring consecutive victories over Grim Reaper, Sir Chromalot, and Ultor. In the semifinal heat, however, Big Brother was eliminated in the opening round by Mace II.

BIG CHEESE

The Grand Frommage took out Shrapnel in the opening round of its heat at the *Robot Wars* 3 (UK) tournament, then eliminated a tough Anorakaphobia in Round 2. Big Cheese was then turned into Swiss Cheese and defeated by the eventual tournament winner, Chaos II, in the heat finals.

BIGGER BROTHER

In the world of boxing, British heavyweights are most often seen horizontal and taking a snooze—but that's not the case when it comes to battling robots. British heavyweight competitor Bigger Brother uses compressed gas to take out its opponents. It comes equipped with a carbon-dioxide-powered ram flipper, which is capable of flipping a foe like a flapjack.

Main builder Ian Watts of Brighton, Sussex, United Kingdom, spent nine months working full-time to build Bigger Brother. He operates and maintains the bot with the help of teammates Wayland Davies and Cath Watts.

At *BattleBots*' Season 2 heavyweight contest, Bigger Brother knocked a tough spinner named Mauler 5150 on its back to advance past the opening round but was eliminated in its next bout by Voltronic.

BIOHAZARD

BioHazard, a heavyweight bot and veteran of multiple *BattleBot* shows, has a lifting arm as its primary weapon. Built by Carlo Bertocchini of

Belmont, California, BioHazard features six-wheel drive and anti-wedge fenders (so that wedge-shaped opponents cannot get underneath it and flip it over).

Since BioHazard is only four inches tall, it has an extremely low center of gravity, which makes it difficult to flip. The bot features no pneumatics or hydraulics, and no welding. BioHazard is held together by over seven hundred machine screws.

In *BattleBots*' Season 2 heavyweight competition, BioHazard flipped Suicidal Tendencies right onto the kill saws to earn a KO victory in the opening round. The bot then defeated Nightmare, M.O.E. and, in the semifinals, Frenzy. In the finals, BioHazard twice flipped the defending champion, Vlad the Impaler, to win the decision and the Giant Nut.

The following season BioHazard came very close to successfully defending its championship. Close, but no cigar, as it earned second place at *BattleBots*' Season 3 tournament. The bot won its first four matches before finally getting bumped out of the competition. BioHazard knocked out Little Sister in the opening round, won a lopsided 42–3 decision over Incoming in the second round, outpointed Tazbot 27–18 in Round 3, and outpointed OverKill 38–7 in the fourth round. In the finals, BioHazard was defeated 29–16 by Son of Whyachi.

BIT RATE

Built by John Kittelsrud and Team Droid, this bot competed in the 2000 BotBash in Phoenix, Arizona. It featured superstructure by Radio Shack (an old R/C car), a stock motor, an out-of-date German seven-channel FM PCM radio-control rig, and a titan speed controller.

The body is composed of vacuum-formed-1⁄16-of-an-inch polystyrene and cardboard from the United States Postal Service. It has a 5-volt power regulator for the receiver and ESC. It is held together with tape and epoxy.

Bit Rate won three out of five fights at the 2000 BotBash. It performed extremely well in the obstacle course, and earned an overall first place.

BLACK OPS

Black Ops was built by Rob Meyer of Tucson, Arizona. It lost in the opening round of *BattleBots*' Season 3 heavyweight tournament, thoroughly chomped by Chewbot.

BLACK WARRIOR

Keegan Hutto and Team Nose Probe of Decatur, Alabama, built Black Warrior, which KO'd General Gau in the opening round of *BattleBots'* Season 3 lightweight tournament, but was carried out on its shield in Round 2, KO'd by The Wacky Compass.

BLADE

Blade performed very well at the *Robot Wars* 3 (UK) tournament. It won its opening heat with consecutive victories over Spike, Agent Orange, and Aggrobot. In the semifinal heat, however, Blade lost in the opening round to Beast of Bodmin.

BLADE RUNNER

Blade Runner spins and whacks its foes across the arena like Barry Bonds laying the wood to a grooved fastball. Actually, Blade Runner is more like Mickey Mantle because not only can it hit home runs, but it can do it from either side of the plate. Blade Runner is a switch hitter and can whack an opponent equally hard—the equivalent of a .45-caliber bullet at point-blank range—no matter which direction it is spinning. Blade Runner cost about $5,000 to build. It was built in about four months by Ilya Polyakov, Ben LaRue, Mike Mass, and Team Carnivore of Dix Hills, New York.

Blade Runner doesn't whack with a bat, however. No blunt trauma here. Blade Runner prefers sharp trauma, and attacks with a thirty-five-pound pickax. According to Polyakov, "The robot is essentially a pickax mounted between two wheels. Its main attack is to rotate the wheels in opposite directions to spin the pickax at eighty-six miles per hour."

In *BattleBots'* Season 1 middleweight tournament, Blade Runner defeated Bad Attitude before being eliminated in the second round by Spin Orbiting Force. It knocked out Instigator in the opening round of *BattleBots'* Season 3 tournament, then was itself KO'd in Round 2 by Little Drummer Boy.

BLENDO

Jamie Hyneman of San Francisco, California, built the heavyweight robot known as Blendo, with the help of teammates Cindy Hobbs, Chris Gaw, Alberto Carreno, Leo Garza, Keegan Hutto, and J. P. Hutto. The bot, which was under construction off and on for four years, has a rotating blade as its primary weapon.

Team Blendo was very disappointed at *BattleBots*' Season 3 heavyweight tournament, as they were eliminated in the opening round, after being knocked unconscious by Bacchus.

BLOOD DRAGON

Malcolm Clark and his band of "Raving Lunatics," all from Long Beach, California, built Blood Dragon, which won a 30–15 decision over Deviation Response in the opening round of *BattleBots*' Season 3 lightweight tournament. It advanced past the second round because of a forfeit, then lost in Round 3 to Dr. Inferno Jr., coming up on the extreme short end of a 40–5 decision.

BLOOD MOON

In *BattleBots*' Season 3 lightweight tournament, Blood Moon—built by Jordan Hackney of San Diego, California—defeated Texas Stampede 27–18, T.U.S.K. 32–13, and Shaft 26–19 before being KO'd in the fourth round by Mouser Mecha-Catbot.

BLUE STREAK

Blue Streak was built by Jason Tajima, Nick Jose, and Mike Tajima of San Diego, California. Built on wheels and battery powered, its primary weapon is a pair of lifting jaws. As an interesting design element, this blue bot comes with neon lights.

BLUNT FORCE TRAUMA

Built on wheels and battery powered, the aptly named Blunt Force Trauma's primary weapon is a "horizontal plane rotating inertia," says main builder Kevin Walsh of San Francisco, California. He and teammates Rand Statman and Kevin Shultis took six weeks to build Blunt Force Trauma.

B.O.B. (BASHER OF BOTS)

Team B.O.B. is headed by Wray Russ of Brentwood, California. In *BattleBots*' Season 3 heavyweight tournament, B.O.B. KO'd Yoshik in the opening round, won a 24–21 decision over Nasty Overbite in Round 2, then lost a 30–15 decision to Omega-13 in the third round.

BODYHAMMER

Bodyhammer was the winner of its heat at the first *Robot Wars* UK series, but lost in the finals to Roadblock.

BOELTER BEAST

Designed to fight like a main battle tank, Boelter Beast has two high-torque wheelchair motors that are used to power the four-wheel drive. Boelter Beast was built with two speed controllers, which gave its operators great maneuverability.

Depending on the opponent, the bot comes with interchangeable wedges, one that enables it to get under opponents, and another that allows it to smash through obstacles. Strong but light, Boelter Beast is built with an aluminum frame and weighs around 160 pounds.

The head builder is Brian Posner, but this was a real team effort. Aiding Posner every step of the way were the members of the American Society of Mechanical Engineers, UCLA division. Posner, who when not designing robots is most apt to be found fine-tuning a motorcycle or a Volkswagen car, is proud to point out that this was the first robot he and the society have ever built.

BOLT FROM THE BLUE

At the *Robot Wars* 4 (UK) tourney, Bolt From The Blue lost in the opening round of its heat to Fire Storm 2.

BOLTZ

Boltz was created by Jason Barger and Brent Hunter of Los Angeles, California, who drew upon their experience in LEGO Mindstorms. Boltz can tow both of its creators around on their Rollerblades and, when it really wants a workout, it can also pull Jason's pickup truck.

Boltz can travel at a top speed of fourteen miles per hour and can spin at 114 revolutions per minute. Its primary weapon is a spiked ram wedge. It has side impact wood framing and a reinforced plexiglass shell.

Its chassis is elevated above the ground to avoid kill saws and lifters. A broad wheel base and shock absorbers help keep Boltz from flipping over.

Boltz was refurbished and competed in England as Mad Cow.
See also MAD COW (II).

BOOBY TRAP

Booby Trap is the creation of Kimberly Everhart and Team Half-Life, Inc., of Mount Airy, Maryland. In *BattleBots*' Season 3 middleweight tournament, Booby Trap knocked out Aftermath in the opening round, won a lopsided 39–6 decision over Cyclonic Plague in Round 2, and lost a 28–17 decision to Bouncing Betty in the third round.

BOTKNOCKER 19

Team Botknocker is headed up by Carl Moyer of Grantville, Pennsylvania. At *BattleBots*' Season 3 heavyweight tournament, Botknocker 19 KO'd Reactore in the opening round, then was defeated by a whopping 43–2 score by Little Sister in the second round.

BOT-A-BING

What's the first bot to wear a pinkie ring? That's an easy one: It's Bot-a-Bing—the creation of Eric Mallory and Cloud Cover Robotics of Houston, Texas. Bot-a-Bing was rubbed out—gangland-style, you might say—in the opening round of *BattleBots*' Season 3 middleweight tournament by Twin Paradox.

BOTILLA THE HUN

Bruce Mobley and Team Mobot of Wilmington, North Carolina, built Botilla the Hun, which came up on the short end of a one-sided battle in the opening round of *BattleBots*' Season 3 lightweight tournament. It was outpointed 39–6 by Carnage Raptor.

BOT MULCHER

Bot Mulcher looks like it can grind up another bot and spit it out, but has lacked success in competition. The invention of Richard Pierce of Las Cruces, New Mexico, Bot Mulcher was masticated by Alabama Slammer in the opening round of *BattleBots*' Season 3 middleweight tournament.

BOTTOM DWELLER

Built by Danis Carter of San Diego, California, Bottom Dweller turned out to live at the bottom of the bot standings as well as it lost in the opening round of *BattleBots*' Season 3 middleweight tournament to Violator.

BOTULISTIC

Jay Johnson of Tallahassee, Florida, is responsible for Botulistic, which lost a 24–21 decision to Trilobot in the opening round of *BattleBots*' Season 3 lightweight tournament.

BOTZILLA

All dressed up and ready to flatten a miniature urbanscape comes this killing machine, which runs on stainless steel tracks and attacks with a powerful pneumatic flipping arm. BotZilla is powered by two Bodine 24-volt, 0.5-horsepower motors and features a full cantilevered live suspension, giving it superior traction and handling on all surfaces.

It was built by Lawrence Feir, age thirty-nine, of Greensboro, North Carolina, who is a professional photographer working in the aerospace industry. His hobbies include sailing R/C yachts, building and tinkering with most anything, and ballooning.

Feir has been tinkering in bots for more than a decade, but his first bot competition came at *Robotica II* in 2001.

BOUNCING BETTY

Conjuring up magical and mystical images of nymphs on trampolines, Bouncing Betty had great success at *BattleBots*' Season 3 middleweight tournament, winning three bouts before elimination. The creation of Henry Hillman Jr. of Portland, Oregon, Bouncing Betty won a 35–10 decision over Trainwreck in the opening round, and a 33–12 decision over Wulfbane in Round 2. In an oddly appropriate matchup, Bouncing Betty evaded the attack of Booby Trap to win a 28–17 decision in the third round, before losing by a KO to Little Drummer Boy.

BRAIN MOLD

Carl Lantz and his team of "Poor College Guys" from Lubbock, Texas, dipped into the beer money to build Brain Mold, which was stomped

34–11 by Recycler in the opening round of *BattleBots'* Season 3 lightweight tournament.

THE BRAINSTERS OF TRISKELLIAN

This short robot with the long name is a superheavyweight built by Francis Ryan, who is not from Triskellian (wherever that is), but rather from Binghamton, New York. This bot was victorious in the opening round of Season 3 of *BattleBots,* taking a 27–18 decision over The Probe. The Brainsters of Triskellian had less luck in the second round, running into the formidable Hammertime and losing by a KO.

BRIMHUR

Brimhur was eliminated in the opening round of the *Robot Wars* 3 (UK) tournament by Weeliwako.

BUDDLESTOMPER

David Hall and the Silicon Valley Destruction Company of San Jose, California, built BuddleStomper, which scored a KO over Chunky 357 in the opening round of *BattleBots'* Season 3 lightweight tournament. In Round 2, however, BuddleStomper lost a 34–11 decision to Crashius Clay.

BUDDY LEE DON'T PLAY IN THE STREET

Buddy Lee, that doll-like demagogue of denim, rides atop this bot that bears his name. Built on wheels and battery powered, Buddy Lee Don't Play in the Street was constructed by Nola Garcia and Mercy Cruz of Miami, Florida. Built for power rather than speed, this middleweight bot claims "pushing power" to be its primary weapon.

BUDDY LEE STAY IN YOUR SEAT

Good thing Buddy Lee is resilient, because his bot was thoroughly thrashed in the opening round of *BattleBots'* Season 3 middleweight tournament by Eraser. Turns out you *can* bust 'em. Another entry from Nola Garcia and Team Fembot of Miami, Florida.

BULLDOG

Built by John Keyes of Woodway, Washington, along with his team, which he calls Computer Keyes, Bulldog knocked out Purple Haze in the first round of *BattleBots*' Season 3 lightweight tournament, then lost by a KO to Whirl Wep X25 in the second round.

BULLDOG BREED

At the *Robot Wars* 3 (UK) competition, Bulldog Breed lost in the opening round of its heat to Robo Pig.

BUMBLE BOT

Bzzzzz, bzzzzz. In an all-insect bout, Bumble Bot defeated Milly Ann Bug in the opening round of the *Robot Wars* 3 (UK) tournament, then was eliminated by Ultor in the second round.

BUMPER BAWT

Eric Hagman of Toronto, Ontario, Canada, constructed Bumper Bawt, which was knocked out in the opening round of *BattleBots*' Season 3 middleweight tournament by T-Minus. Bumper Bawt is built on wheels and runs on batteries. Its primary weapon is a battery-powered, 1,500-pound pneumatic ram.

This bot is constructed entirely of modular components, which greatly cuts down on repair time. Its armor, motors, controllers, and fuse can all be replaced in a few minutes. Not including the conceptualization and design time, Bumper Bawt took about twenty workdays to build.

C

CALIBAN

Named after an earthy character in *The Tempest*—oooh, Shakespeare!—Caliban was built by David Norris and his decidedly Shakespearean crew, Team WhoopAss, from Redmond, Washington. Caliban was banished to the underworld in the opening round of *BattleBots*' Season 3 middleweight tournament by SMD.

CANNIBOT

A superheavyweight out of Austin, Texas, Cannibot was built by James Miller. Cannibot never got an opportunity to eat its own kind, as it was eliminated in the first round of *BattleBots*' Season 3, losing to Africanized Killer Bot by a score of 27–18.

CAN OPENER

Can Opener was constructed by Adam Baxter and Team Annihilator of Ormond Beach, Florida. It defeated Spin Cycle by a whopping score of 44–1 in the opening round of *BattleBots*' Season 3 middleweight tournament, then lost a close 24–21 decision to Cheap Shot in Round 2.

CARNAGE RAPTOR

Team Raptor's lightweight entry, Carnage Raptor won three bouts in *BattleBots*' Season 3 lightweight tournament before it was eliminated. Built by Robert Pitzer of Chandler, Arizona, Carnage Raptor won a 38–7 decision in the mud, the blood, and the beer over A Bot Named Sue in Round 1, a 39–6 decision over Botilla the Hun in Round 2, knocked out Malice in Round 3, and was KO'd by Sallad in Round 4.

CASSIUS II

At the *Robot Wars* 3 (UK) competition, Cassius II defeated Dundee in the opening round of its heat, then lost in the second round to Pussycat. Boxing fans will find an interesting coincidence in the first matchup, since Angelo Dundee was trainer to Cassius Clay, the boxer who became Muhammad Ali.

CENTER PUNCH

Center Punch was constructed by Team Green Iguana of Lake Helen, Florida, under the supervision of Head Iguana Tom "Angry" Beaver. Center Punch did all right in *BattleBots*' Season 3 heavyweight tournament, winning its opening-round match by a score of 28–17 over Tool Chop, and its second-round match 29–16 over Incisor. Center Punch lost in Round 3, however, when it was KO'd by KillerHurtz.

CENTIPEDE

What's the best town for building bots? Here's my nomination: Brainerd, Minnesota, where William Arden and Myriapod built Centipede, which lost a 26–19 decision to Herr Gepoünden in the first round of *BattleBots*' Season 3 lightweight tournament.

CENTURION

At the *Robot Wars* 3 (UK) competition, Centurion defeated Excaliber in the opening round of its heat, then was eliminated in the second round by that silly-millimeter-longer bot, 101.

CERBERUS

At *BattleBots*' Season 3 middleweight tournament, Cerberus defeated Nibbler, won a 29–19 decision over Travesty, then was knocked out by Double Agent in Round 3. At the *Robot Wars* 3 (UK) competition, this bot made it all the way to the finals of its heat, defeating KillerHurtz and Griffon before being eliminated by heat winner Thing II. Cerberus was built by Robert Davis and Team Hells Mechanics of Westfield, New Jersey.

CHALLENGER II

At the *Robot Wars 3* (UK) competition, Challenger II defeated Atlas in the opening round of its heat, but then was eliminated in the second round by Panzer.

CHAMELEON

Built by Greg Hjelstrom of Las Vegas, Nevada, Chameleon lost a 28–17 decision in the opening round of *BattleBots*' Season 3 middleweight tournament to Exodus 2001.

CHAOS II

Chaos II is powered by two Bosch 400-watt, 12-volt motors running at 24 volts for extra power. The weapon is a high-pressure carbon-dioxide-powered forward-facing flip-up. The chassis is 10-millimeter Dural plate with polycarbonate covers on top. The flip-up will self-right the robot if it gets turned over.

Chaos II was winner of the *Robot Wars* 3 (UK) tournament. Chaos II won its heat with consecutive victories over Crocodilotron, Sonic, and Big Cheese. In its semifinal heat, it defeated Trident, Mace II, and Fire Storm. In the tournament's final match it met and vanquished Hypnodisc. At *Robot Wars* 4 (UK), Chaos II won its opening heat with victories over Medusa 2000 and Atomic.

CHEAP SHOT

Nathan Roseborrough and his crew, known as the Poor College Guys, all of Lubbock, Texas, built Cheap Shot, which knocked out Palindrome in the opening round of *BattleBots'* Season 3 middleweight tournament, outpointed Can Opener in Round 2 by a score of 24–21, then was KO'd by T-Minus in the third round.

CHEWBOT

Built by Larry Beckman of Lake Park, Florida, Chewbot made it to the second round of *BattleBots'* Season 3 heavyweight tournament, knocking out Black Ops in the opener, then losing a 35–10 decision to Kritikal Mass II.

CHIA-BOT

Just add water. It grows! Chia-bot was constructed by Will Wright and the Robot Action League of Orinda, California. It was given a good haircut, then knocked cold in the opening round of *BattleBots'* Season 3 middleweight tournament by Village Idiot.

CHIMERA

Tom Hall and AVHS Robotics of Pleasanton, California, constructed Chimera, which lost a 32–13 decision to The Archduke in the opening round of *BattleBots'* Season 3 lightweight tournament.

CHIN-KILLA

Built to look like late-night talk-show host Jay Leno. Chin-Killa has never officially competed in a TV tournament, but it did fight in an exhibition match during *BattleBots'* Season 2.

In that bout, Chin-Killa used "The Power of the Chin" to flip a bot named Ginsu. Leno and his chin were there in person for the exhibition, cheering on his mechanical likeness.

CHOMP

Chomp, another bot out of the PacMan school of robot building, was constructed by Kevin Lung (leader of Team LungFish) from Sunnyvale, California. Chomp acquired lockjaw in its match against I-Beam in the opening round of *BattleBots*' Season 3 heavyweight tournament, and was KO'd.

CHUNKY 357

John Ghormley and The Chunky Monkeys of Rancho Palos Verdes, California, created Chunky 357, which was KO'd in the opening round of *BattleBots*' Season 3 lightweight tournament.

CIRCUIT BREAKER

Although a heavyweight, Circuit Breaker depends on quickness to defeat its opponents—quickness and the disabling thumping of its pounding hammer. Built on wheels and battery powered, the bot is the creation of Josh O'Briant of San Jose, California, who built Circuit Breaker with the help of teammates Dave Fieber and Kenneth Ton over "three months and many hours." In the opening round of *BattleBots*' Season 3 heavyweight tournament, Circuit Breaker was KO'd by FrostBite.

CLOSE ENOUGH

David Gilson, Brenda Gilson, and Elwin McKernan of Omaha, Nebraska, constructed the 220-pound monster known as Close Enough. The bot operates on two-cycle gas motors. Its strength is its low ground clearance. Its weakness is the ineffectiveness in combat of its primary weapon, which is comprised of a set of ramming teeth. In the "Robot Rebellion" episode of *Robot Wars: Extreme Warriors,* in which entrants competed for the right to take on the fearsome *Robot Wars*' house bots, Close Enough was defeated by Silver Box in the eliminator round. At the end of the match, Close Enough found itself airborne after being pushed onto the arena's built-in flipper, then on fire as *Robot Wars*' house bot Shunt pushed it into the barbeque pit.

COLD STEEL

Frank Harris and Team Tarantula of Tujunga, California, built Cold Steel, which was knocked out in the opening round of *BattleBots*' Season 3 lightweight tournament by Fang.

COLLATERAL DAMAGE

This bot has the same name as an Arnold Schwarzenegger film—and the bot came first! Collateral Damage was built by Mike Pelstring and Team Kinetic Energy of Santa Cruz, California. Collateral Damage lost in the opening round of *BattleBots*' Season 3 heavyweight tournament to Crash Test Dummy in a 26–19 decision.

COMPLETE CONTROL

Built by Derek Young and Team Automatum of Burnaby, British Columbia, Canada, Complete Control knocked out a bot named Subject to Change Without Reason in the opening round of *BattleBots*' Season 3 middleweight tournament, then lost via KO to F5 in Round 2. Complete Control is built on wheels and battery powered. Its primary weapon, a lifting claw, is powered both by a battery and compressed carbon dioxide. By using a pneumatic grabbing arm and an electric lifting fork, Complete Control can grab and lift an opponent, and carry them to their doom.

CORPORAL PUNISHMENT

At the *Robot Wars* 3 (UK) competition, Corporal Punishment was eliminated in the opening round of its heat by the jawesome Ally Gator.

COUNT BOTULA

Count Botula was built by Alan Martinson of Austin, Texas, who always tries to keep his bot out of the sun. Unfortunately, the Count had a stake driven through its heart by Spare Parts in the opening round of *BattleBots*' Season 3 middleweight tournament.

CRAB MEAT

Built by Don Harper of Orlando, Florida, Crab Meat knocked out My Son in the first round of *BattleBots*' Season 3 heavyweight tournament, then lost via KO in Round 2 to Son of Whyachi.

CRASHER KNASHER

At the *Robot Wars* 3 (UK) tournament, Crasher Knasher was eliminated in the opening round of its heat by the eventual heat winner, Fire Storm.

CRASHIUS CLAY

Floats like a butterfly and stings like a bee! Built by Steven DeZwart and Team Green Machine of Newport Beach, California, Crashius Clay won a pair of decisions before it was eliminated from *BattleBots*' Season 3 lightweight tournament. It outpointed BuddleStomper 34–11 and Vermicious Kenid 30–15 before coming up on the short end of a 33–12 decision to Skid Mark.

CRASH TEST DUMMY

Take a poll of everyone's favorite name for a warrior bot and this machine's monicker would be sure to come up near the top. Crash Test Dummy was constructed by Robert Doerr of St. Clair Shores, Michigan. In *BattleBots*' Season 3 heavyweight tournament, Crash Test Dummy defeated Collateral Damage in the opening round by a 26–19 score, then lost in Round 2 to BattleRat by a score of 33–12.

THE CREATURE

From the Black Lagoon to the *Robot Wars*' arena comes The Creature, which defeated Invertibrat in the opening round of its heat at the *Robot Wars* 4 (UK) tournament. It lost to Gemini in the second round.

CRIPPLER

At the *Robot Wars* 3 (UK) competition, Crippler became the cripplee and was eliminated in the opening round of its heat by All Torque.

CRISIS MANAGEMENT

Crisis Management was built by J. D. Streett, age thirty-eight, Karl Nettmann, age thirty, and Charles Coolie, age thirty, of Los Angeles and Glendale, California. Their first major competition was Robot Wars '95, at which they were a quarterfinalist and winner of the strangest entry.

Among Crisis Management's strong points are its low center of gravity and the fact that it is invertible. It has four-wheel drive and attacks with a cable-cutting ankle biter. The Crisis Management team, when they are not building pugilistic androids, work as special effects coordinators in Tinsel Town.

In *Robotica*'s Season 1 tournament, Crisis Management lost in the preliminary rounds to Jawbreaker.

CRITICAL CONDITION

Critical Condition lost in the opening round of *BattleBots*' Season 3 lightweight tournament to Spin Reaper by a 25–20 score. Critical Condition was built by Robert Moore and Team Ares of Dunedin, Florida.

CROCODILOTRON

Crocodilotron might have frightened Captain Hook, but it didn't intimidate Chaos II. The eventual tournament winner eliminated Crocodilotron in the opening round at *Robot Wars* 3 (UK).

CRONOS

At the *Robot Wars* 4 (UK) tourney, Cronos lost in the opening round of its heat to Crusader 2.

CROUCHING PUPPY, HIDDEN KITTEN

The brainchild of Shaun Fetherston and Team CPHK of Portland, Oregon, Crouching Puppy, Hidden Kitten effected a soaring victory over Mordicus in the opening round of *BattleBots*' Season 3 middleweight tournament, then lost a crushing 42–3 decision to Uplifting Experience in Round 2.

CRUSADER 2

At the *Robot Wars* 4 (UK) tourney, Crusader 2 defeated Cronos in the opening round of its heat, then was defeated by Mortis.

CRUSHER

At the *Robot Wars* 3 (UK) competition, Crusher lost in the opening round of its heat to Beast of Bodmin.

THE CRUSHER

The Crusher was built by Thomas Petruccelli and Team Attitude of San Jose, California. At *BattleBots*' Season 1 lightweight tournament, The Crusher made it to the second round before it was eliminated by the eventual winner, Backlash—although the truth was that The Crusher, which is built on wheels and battery powered, lost as much to the BattleBox's kill saws as it did to its opponent. It was a KO victim of Herr Gepoünden in the first round of *BattleBots*' Season 3 lightweight tournament.

CUAD THE CRUSHER

Cuad is a superheavyweight built by Matt Ulrey of Plain City, Ohio. During the *BattleBots*' Season 3 tournament, Cuad defeated Jack the Gripper by KO in the opening round, followed that up with a decisive 32–13 victory over Overlord in the second round, but then met its match in the third round when it lost by a hair to Techno Destructo 23–22.

EL CUCUY

Gerardo Serafin and Nexus 6 of Gardena, California, are responsible for El Cucuy, which was knocked out by Scheduled for Destruction in the opening round of *BattleBots*' Season 3 middleweight tourney.

CUNNING PLAN

Cunning Plan was the winner of its heat at the first *Robot Wars* UK series, but lost in the finals to Roadblock.

CYCLONE

Built by Kevin Knoedler of Newbury Park, California, Cyclone is a superheavyweight bot that was defeated in the opening round of *BattleBots*' Season 3 tourney. The defeat was dished out by OJ, who won a 30–15 decision.

THE CYCLONIC PLAGUE

The Cyclonic Plague was pretty doggone sick itself as it lost a 39–6 decision in the opening round of *BattleBots*' Season 3 middleweight tournament to Booby Trap. This sick puppy was the creation of Jonathan Grimm and Team G.R.U.N.T. of Corona, California.

D

DAISY

Daisy, constructed by Derek Zahn of Boulder, Colorado, KO'd Godtilla in the first round of *BattleBots*' Season 3 lightweight tournament, then lost 34–11 in Round 2 to Reaper. At the *Robot Wars* 3 (UK) tournament, Daisy lost in the opening round to Aggrobot.

DAISY CHOPPER

At the *Robot Wars* 3 (UK) competition, Daisy Chopper was defeated by Griffon in the opening round of its heat.

DARKE DESTROYER

At the *Robot Wars* 3 (UK) competition, Darke Destroyer did very well, advancing to the finals of its heat before being eliminated. This bot whupped Sgt. Meikle and T II before losing to Gravedigger.

DARK MATTER

Dark Matter was nearly skunked, defeated by a score of 39–6 by Wedge of Doom in the opening round of *BattleBots*' Season 3 lightweight tournament. Built by Ted Zeiger and Team Cosmos of Boulder, Colorado.

DARK STEEL

Russ Barrow and Team Knowledge of Richardson, Texas, came up with Dark Steel, a bot that was KO'd in the opening round of *BattleBots*' Season 3 lightweight tournament by Junkyard Dog.

DAS BOT

Built on wheels and battery powered, Das Bot has a pneumatic lifting arm that operates on compressed carbon dioxide as its main weapon. It was built by Paul Mathus and Matt Bazeley of Oakland, California. Das Bot is a powerful lightweight and can achieve up to 1 horsepower of high-torque drivetrain power. It has four-wheel drive and gyro-stabilized steering. It has a heat-treated alloy steel frame and shock-mounted double-layer Lexan armor. At *BattleBots*' Season 1 lightweight tournament, Das Bot defeated Rott-Bot 2000, Shrike, and Ziggo before being annihilated (we're talking mass destruction here) by eventual winner Backlash.

DAWN OF DESTRUCTION

Dawn of Destruction, a superheavyweight, was built by Jim Klein of Columbiana, Ohio. This bot had a great deal of success at *BattleBots*' Season 3 tournament, winning three consecutive decisions—over The Swarm by a score of 23–22, over Gyrax 24–21, and over Eradicator 33–12. Dawn of Destructions' luck then ran out as it was KO'd in the fourth round by Diesector.

DEADBLOW

Grant Imahara, Preston Donovan, and Nelson Hall of Oakland, California, created Deadblow, which made it to the finals in the middleweight division at *BattleBots*' Season 1 tournament before losing to Hazard. At that competition, Deadblow was given the Comeback Robot of the Year award. It earned that title during a match against Alien Gladiator when its swinging titanium spiked hammer, which operates on compressed carbon dioxide, snapped off due to metal fatigue. Without missing a beat, Deadblow continued its attack as a ramming bot.

It knocked out Spinster in the first round of *BattleBots*' Season 3 tournament, then was itself KO'd by Twin Paradox in Round 2.

DEAD METAL

This house bot on *Robot Wars: Extreme Warriors* weighs 250 pounds and measures sixty-four inches by forty inches by twenty-eight inches. It gets its juice from dual electric motors and has two weapons: a pincer, and a saw blade that turns at an unbelievable 3,000 revolutions per minute. Stick your hand in there and they'll be calling you Lefty!

DEATH BY MONKEYS

Great name. Great bot! Death by Monkeys scored four consecutive knockout victories at *BattleBots*' Season 3 lightweight tournament before being eliminated via KO. Its victims were, in order, Weed Wacker, Unlicensed Chiropractor, Green Dragon, and Sisyphus. It was then victimized by Herr Gepoünden.

DEATH WARMED UP

At the *Robot Wars* 3 (UK) competition, Death Warmed Up lost in the opening round of its heat to Stealth.

DEATHWISHBONE

Built by Chad Castillo and a troupe he calls Tesla's Stooges, of Oakland, California, Deathwishbone lost in the opening round of *BattleBots'* Season 3 heavyweight tournament, KO'd by Gang Green Weasel.

DEB BOT

Don't let the name fool ya. Deb Bot may sound like it has nothing more on its mind than a prom or maybe a sweet sixteen, but the truth of the matter is that this bot "thinks" predominantly of death and dismemberment!

A wedge-shaped lovely ready for her social debut, Deb Bot uses twenty-four nicad cells to power the two Astroflight Cobalt 90 motors through a Vantec 36E speed controller. She runs on four wheels and has "tank" steering for directional control.

Deb Bot's lifting arm is powered by a 250-pound linear actuator that is overpowered on 18 volts. Made predominantly of 6061 aluminum with a tube frame, ⅛-inch armor on the sides and ¼-inch armor on the wedge and bottom.

This princess of the ball tips the scales at 95 pounds, yet she has a little oomph behind her. When she smacks you, you know you've been smacked. She's also known for her agility. Hit and move, that's the game plan.

Deb Bot was built by painting contractor Jeff Vasquez, age forty, of Oak Park, California, who has been working in robotics for less than two years. Obviously, he is a quick learner. His hobby is racing radio-controlled electric boats. The fast ones.

Jeff's teammate is David Carriker.

The bot won a 35–10 decision to Narwal in the opening round of *BattleBots'* Season 3 middleweight tournament, won a 24–21 decision over Spare Parts in Season 2, then was KO'd and eliminated by Village Idiot in the third round.

When last seen, Deb Bot was heading to the *Robotica II* tournament.

DECIMATOR

Decimator, a heavyweight veteran of *BattleBots,* can operate both on its belly or its back, so it is flip-proof. To fight Decimator, you had better be flip-proof as well, because the bot's primary weapon is a flipper arm that operates with compressed carbon dioxide.

Getting under Decimator's skin is no easy task, as the bot is guarded by one-half-inch bulletproof Lexan armor. The battery power can generate 17.2 horsepower and Decimator has attained a top speed between fifty and sixty miles per hour.

Decimator was built by Kyle Vogt and Bryan Boettcher of Shawnee, Kansas, over a two-and-a-half-year period, with the help of teammates Charles Vogt and Mark Boettcher.

DELTA II

Built on wheels and battery powered, Delta II has a reciprocating ram as its primary weapon. It was built in four weeks by Mark Supal, David Yruma, Erica Sachs, and Wayne Penn of Warren, Michigan.

DESTRUCTOBOT/WHIRRLING WILLY

This battery-powered and donut-shaped bot, built on wheels, is so nice that they named it twice—they being Keith Van Sickle, Rigoberto Rubio, Andrew Barlow, and Miles Barlow of Ramona, California.

This bot, which took six weeks to build, has been known to go into battle with an onboard computer. It uses both its own spinning and a spring-loaded hammer as its weapons.

DEVIATION RESPONSE

Dave Morris and Team Polafia of Simi Valley, California, built Deviation Response, which lost a 30–15 decision to Blood Dragon in the opening round of *BattleBots*' Season 3 lightweight tournament.

EL DIABLO

A middleweight built on wheels and battery powered, El Diablo (Spanish for The Devil) was built over a two-month period by Zachary Bieber, Robert Marino, and Heather Kyseth of Monrovia, California.

The primary weapon is a swappable trench saw with a large-toothed steel drum. El Diablo has a highly maneuverable track-drive platform. It also has a lifting and dragging tail and an audible and visual FX generator to add excitement to the fight.

So El Diablo not only shows up to fight, it brings its own light and sound show, to boot.

EL DIABLO GRANDE

Like the bot above, El Diablo Grande was built by Zack Bieber and Team Diablo of Monrovia, California. El Diablo Grande was looking to beat the devil out of some bot in the opening round of *BattleBots*' Season 3 heavyweight tournament, but instead lost a close 23–22 decision to Mechavore.

DIAMONDBACK

Jason Stokes of Tewksbury, Massachusetts, came up with Diamond-Back, which was a KO victim of Junkyard Dog in the opening round of *BattleBots*' Season 3 lightweight tournament.

DIE FLEDERMAUS

Blake Mitchell of San Francisco, California, and his troupe called Emergency Third Rail Power Trip built Die Fledermaus. The bot defeated Agsma 26–19 in the opening round of *BattleBots*' Season 3 middleweight tourney, then lost 33–12 to Ice Cube in the second round.

DIESECTOR

Diesector can cut apart things that are a lot tougher than a pickled frog. This battery-powered superheavyweight boasts 360 degrees of offense and defense. It has a customized pickax weapon.

Diesector was built by Donald Hutson of San Diego, California. His teammates are David Cook, David Blakey, Dawn Bernstein, Paul Remond, and Mike Moore.

This bot was the winner of the Giant Nut at *BattleBots*' Season 2 super-heavyweight competition. But winning the nut was not easy. The bot almost didn't advance past the opening round. Sure, Diesector defeated World Peace in that bout, but it was left smoking and in need of extensive repairs before it could meet the bell for the second round. Diesector defeated Rammstein in the second round, pacified War Machine in the semifinals, and defeated Atomic Wedgie in the finals.

At *BattleBots*' Season 3, Diesector picked up right where it left off the previous season and scored a series of impressive knockout victories. It defeated Scrap Daddy's Turbulance in the first round, Dawn of Destruction in the second, and Rammstein in the third. Diesector then lost in the fourth round in a close decision to Vladiator, outpointed 24–21.

DIOTOIR

At the *Robot Wars* 3 (UK) tournament, Diotoir defeated Sting II in the opening match of its heat, then defeated Steel Avenger in the second round. Diotoir was eliminated in the heat finals by Fire Storm. At the *Robot Wars* 4 (UK) tourney, it was eliminated in its first bout.

DISCO COW 3000

It sort of reminds me of the old KC and the Sunshine Band disco classic: *Shake, shake, shake. Shake, shake, shake. Shake your udder, shake your udd-er!* Disco Cow 3000 is the braincalf of Patrick Dwight of Anaheim, California. The cow lost by KO in the opening round of *BattleBots*' Season 3 heavyweight tournament to that mechanical mutant Junkyard Offspring.

DISCO INFERNO

At the *Robot Wars* 4 (UK) tourney, Disco Inferno lost in the opening round of its heat to the eventual heat winner, Dominator 2.

DISPOSABLE HERO

Disposable Hero was eliminated in the opening round of *BattleBots*' Season 1 lightweight tournament by the eventual winner, Backlash. Disposable Hero lacked armor and was easy fodder for Backlash's jaw.

DISPOSE-ALL

Spawned from Brian Nave of Ormond Beach, Florida, and Team Logicom, Dispose-All was disposed of in the opening round of *BattleBots*' Season 3 middleweight tourney by Uplifting Experience. The score: 37–8.

DR. INFERNO JR.

At *BattleBots'* Season 1 lightweight tournament, this bot, built on wheels and battery powered, lost an opening-round match to Alpha Raptor, a bot that eventually made it to that season's finals.

In *BattleBots'* Season 3 lightweight tournament, Dr. Inferno Jr. scored a 40–5 win over Blood Dragon, a 23–22 decision over Toe Crusher, a KO victory over Bad Habit, and a 27–18 decision over Herr Gepoünden.

In the semifinals, it defeated Sallad 36–9 when Sallad became caught under the BattleBox spikes. In the finals, it won a controversial 32–13 decision over Gamma Raptor to win the Giant Nut. The crowd saw it the other way and booed the decision.

It was built over the course of four and a half months by Jason Dante Bardis, Lauren Herold, Mike Wrobleski, and The Infernolab of Santa Barbara, California.

Bardis is a graduate student at the University of California in Santa Barbara, where he is chasing a doctorate degree. Dr. Inferno Jr. is Bardis's ninth bot.

DOMINATOR 2

At the *Robot Wars* 4 (UK) tourney, though unseeded, Dominator 2 won its heat with victories over Henry 2, Shadow of Napalm, and 101.

DOOALL

DooAll, a treaded robot from the superheavyweight division, was built by Scott LaValley of San Rafael, California. His teammates are Trevor Bommersbach and Angelo Figallo. DooAll smashes its opponents with its ram rod, which operates with compressed gas. In *BattleBots'* Season 1 superheavyweight tournament, DooAll made it all the way to the finals before it was defeated by Minion. At the Season 2 competition, DooAll made it as far as the third round and was only eliminated in the closest of decisions by War Machine. At *BattleBots'* Season 3 tournament, DooAll was eliminated in the opening round when it lost a close decision to Anubis, 24–21.

DOOM OF BABYLON

Doom of Babylon is a battering ram of a bot that uses, according to builder Todd Mendenhall of Rancho Palos Verdes, California, "kinetic energy" as its primary weapon. In other words, ram power! The bot, which took seven weeks to build, thrives on power, power, power. Its battery-powered "boost" motors can generate up to 20 horsepower.

THE DOOMSDAY MACHINE

Terry Wynn and The Doom Crew of Troy, Michigan, fathered The Doomsday Machine, which lost a 36–9 decision in the first round of *BattleBots'* Season 3 lightweight tournament to Herr Gepoünden.

DOORSTOP

As the name implies, Doorstop is a wedge-shaped bot, which operates on wheels and under battery power. It is symetrically built and can run with either side up. It was built by David Pabian, Gary Avery, and Devin Ramkin of Phoenix, Arizona.

DOUBLE AGENT

Robert Sica and his self-proclaimed Diginati of Dracut, Massachusetts, devised Double Agent, a highly successful bot that won four bouts and lost one in *BattleBots'* Season 3 middleweight tourney.

Double Agent knocked out Armadillo in the opening round, defeated Scheduled for Destruction in Round 2 by a score of 24–21, KO'd Cerberus in Round 3 and Turbo in Round 4, before losing a 34–11 decision to Sabotage.

DOUBLECROSS

This lightweight veteran of the *BattleBots* tournaments is built on wheels and battery powered. Its primary weapons are two spinning blades, which also operate on batteries. It was built in "about a month" by Andrew Schwirian.

DOUBLE SECRET PROBATION

Bryson DeJong and Team Hamncheez of Lodi, California, built Double Secret Probation, which was outpointed 34–11 by Matts Bammer in the first round of *BattleBots'* Season 3 lightweight tournament.

DRAC'S REVENGE

Drac's Revenge lost to Steg-a-Saw-Us in the opening round of its heat at the *Robot Wars* 3 (UK) competition.

DREADBOT

Dreadbot might sound like it comes from Kingston, Jamaica, where dreadlocks are the hairstyle of choice, but it actually comes from San Jose, California, where it was built over an eight-week period by David Hall, with the help of teammates Roger Baker, Chris Kallai, and John Brownson. Dreadbot's special features include a gyroscopically stabilized control system. That, and its compressed-gas pneumatic ram weapon, inspire dread in all of its opponents.

At *BattleBots*' Season 2 heavyweight competition, Dreadbot lost in the opening round to Voltronic. In the opening round of BattleBots' Season 3 tournament, Dreadbot lost by a KO to Mauler 5150.

DREADNOUGHT

A superheavyweight built by Christian Carlberg of Santa Monica, California, Dreadnought lost to OJ by a 24–21 decision in the opening round of the *BattleBots*' Season 3 tournament. Carlberg calls Dreadnought a "double threat robot, a wedge on one end and a spinning blade on the other."

Across the pond, Dreadnought advanced as far as the finals of its eight-bot heat at the *Robot Wars* 3 (UK) tournament. It defeated Psycho Killer and Rattus Rattus before succumbing in the third round to heat winner Trident.

DREDGE

Built by Gene Burnett of Springfield, Illinois, Dredge was eliminated by Fork-N-Stein in the first round of *BattleBots*' Season 3 heavyweight tournament.

DRILL-O-DILLO

Drill-o-dillo's primary weapon is a rotating six-inch horizontally mounted drill bit in the front. Its secondary weapon is its wedge-shaped rear end. The robot also features twin 0.75-horsepower electric PMDC motors, each 24 volts. Drill-o-dillo uses high-current sealed lead-acid batteries. From a defensive vantage, this robot is difficult to attack because of its ability to turn quickly. Andrew Gardner of Goleta, California, along with teammate Scott Gary, built Drill-o-dillo over about a 200-hour period.

DRILLZILLA

With Tut Tut, Drillzilla won the Tag Team Terror Championship at the *Robot Wars: Extreme Warriors* Season 1 tournament. On their way to victory, this team defeated Rosie the Riveter and Tricerabots, and then General Chompsalot and Run Away.

E

ELECTRIC LUNCH

The creation of Steven Nelson of Sparks, Nevada, Electric Lunch is a superheavyweight that performed very well in the *BattleBots*' Season 3 tournament. In the opening round it defeated Jaws of Death by a 30–15 decision. In the second round it ousted Ronin from the tournament with a KO victory. It then outpointed Odin II 29–19 in the third round before losing a 24–21 decision to Minion in the fourth round.

THE EMASCULATOR

A middleweight veteran of *BattleBots,* The Emasculator's primary weapon is a 60-pound, six-foot-long twin helicopter blade, the "scary-violent blade o' death." The blades are named Cylla and Charybdis. The bot's modular design allows for easy repair. It was built over a six-month span by Ray Scully and Joshua Palmer of San Anselmo, California.

ENDOTHERM

At the *BattleBots*' Season 1 lightweight tournament, Endotherm put up a good fight but was eliminated in the third round via decision by the eventual winner, Backlash.

ENFORCER

Enforcer is a lightweight veteran of *BattleBots*. Its primary weapon is a lifter/saw arm. Both the bot and the weapon are battery powered. That saw-mounted arm can be raised and lowered, or can be used to lift an opponent. The arm has twenty-seven inches of lift when fully extended. It has four-wheel drive and is made out of aluminum. It was constructed off and on over a two-year span by Al Kindle, Chris Cardone, and John Wolan of Edison, New Jersey.

ENUMSCRATCH

This bot was eliminated from *BattleBots*' Season 3 middleweight tourney in the opening round by Alabama Slammer, coming out on the short end of a 37–8 decision. If you can't figure out what the heck the name of this bot means, here is a clue: EnumScratch was built by Brian Patrick of Enumclaw, Washington.

ERADICATOR

Eradicator is a superheavyweight bot built by Ken Gentry of Orlando, Florida. In competition this bot apparently doesn't believe in a close-fought battle. In *BattleBots*' Season 3 tournament, Eradicator first defeated Little Egypt 39–6, then was just as soundly thrashed by Dawn of Destruction in the second round, losing 33–12.

ERASER

One of the toughest middleweight bots to appear at the *BattleBots*' Season 3 tourney, Eraser was constructed by Kenneth Fuiks and his Pencil Neck Geeks of Fresno, California. In order, Eraser defeated Rebob (by KO), B.E.H.EMOTH (28–17), Horrifica (KO), Buddy Lee Stay in Your Seat (KO), and Village Idiot (KO). Only in Round 6, the finals, did Eraser succumb—KO'd by Little Drummer Boy.

ERIC

At the *Robot Wars* 3 (UK) competition, Eric was eliminated in the opening round of its heat by King Buxton.

ERIC THE RED

Eric the Red charged horns-first into defeat in the first round of *BattleBots*' Season 3 lightweight tournament, losing a 37–8 decision to Wacker. Built by Eric Dalnes and Team Viking USA of Santa Ana, California.

EVIL BEAVER

An old 486 computer with all of its internal electronics removed forms the body of the battling robot known as Evil Beaver. It is powered by a 24-volt gel cell–type battery and uses surplus 12-volt DC motors with a built-in gearbox with gear reduction and lightweight ABS plastic wheels.

The chain saw, which comprises Evil Beaver's main weapon, is a Black & Decker small electric. If Evil Beaver's chain saw doesn't get you, its drill will. The drill on the front is a hacked cordless screwdriver with a steel carbide tip.

The nose part of the robot is actually a heating-duct vent adapter. The robot is controlled with a homemade speed controller that uses relays to hit micro switches to send current to the motors. The radio is a Futaba six-channel radio with ground crystals.

Evil Beaver's builders are fourteen-year-old Kyle Rader of Saratoga, California, and Charles Rader of Camp Peavy, Palo Alto, California. The pair had a combined five years of robot-building experience before appearing on *BattleBots* and had previously earned a fourth place in the San Francisco Robotics Sumo Contest.

In *Robotica*'s Season 1 tournament, Evil Beaver was eliminated in the preliminary rounds by Viper.

EVIL CON CARNE

Evil Con Carne is the hellish result of work by Terry Talton and Team DeSade of Dallas, Georgia. In *BattleBots*' Season 3 middleweight tourney, Evil Con Carne defeated Outlaw by a score of 33–12, KO'd Gungnire, then lost to Alabama Slammer in the third round by a score of 37–8.

EVIL FISH TANK

Yes, it's the dawning of the Age of Aquariums, and Evil Fish Tank is on the prowl. As its name implies, it has a see-through body so you can see all its innards. Its primary weapon is a compressed-gas-powered actuated titanium ramp.

It was built by Dan Danknick and Tony Buchignani of Orange, California. At *BattleBots*' Season 2 lightweight competition, this bot made it as far as the second round, when it was eliminated by Toe Crusher.

EVIL WEEVIL

My guess is that, if ever a bot flies over the Grand Canyon on a motorcycle, Evil Weevil will be the one to do it. At the *Robot Wars* 3 (UK) competition, Evil Weevil won its heat, scoring consecutive victories over Triterobot, Flip Flop Fly, and Panzer. In the semifinal heat, Evil Weevil lost in the opening round to the eventual second-place winner, Hypnodisc.

EXCALIBER

Excalibur is the name of the sword that King Arthur pulled out of the stone. Unfortunately, there was no pulling this bot, named Excaliber, out of the fire as it was eliminated in the opening round of its heat at the *Robot Wars* 3 (UK) competition by Centurion.

EXODUS 2001

Patrick Verner of Comstock Park, Michigan, is responsible for this winner, which won three bouts before losing in *BattleBots*' Season 3 middleweight tourney. Exodus 2001 defeated Meltdown 32–13, Chameleon 28–17, and Short Order Cheff 27–18. In Round 4 Exodus 2001 was knocked out by Zion.

EYE OF THE STORM

Isn't there supposed to be calm at the eye of the storm? Well, not in this case. This bot was built by Jonathan Bedford of Lawndale, California. It defeated Maneater by a score of 25–20 in the opening round of *BattleBots*' Season 3 middleweight tourney, then lost in Round 2 to Village Idiot, coming up on the short end of a 30–15 decision.

F

F5

F5 was one of the best middleweight bots to compete in San Francisco at the Season 3 tournament. The bot made it into the sixth round before it was eliminated. After scoring a 28–17 decision over Maggot in the opener, F5 scored four consecutive KOs, over (in order) Li'l Chipper, Violator, Ankle Biter, and Complete Control. In Round 6, F5 was KO'd by the eventual winner, Hazard. The bot was built by Dan Bovinich of Roseville, Michigan.

FACET

At the *Robot Wars* 3 (UK) tournament, Facet defeated Terminal Ferocity in the opening round of its heat, but was eliminated in the second round by the eventual heat winner, Fire Storm.

FALCON

Built by David Temple, Rand Whillock, and Mark Whillock, Falcon weighs 214 pounds and measures fifty-four inches by thirty-six inches by seventeen inches. It is powered by a pair of 12-volt motors and its main weapon is a lifting spike.

Falcon participated, finishing last, in the Tag Team Terror tournament during Season 1 of *Robot Wars: Extreme Warriors.* It teamed with Joker, and together they lost first to the team of General Chompsalot and Run Away, and then, in a battle of the losers, to Rosie the Riveter and Tricerabot.

On the "Civil War" episode during that same season, Falcon represented the North and was eliminated in the opening round by The Six Million Dollar Mouse.

FANG

This bot was built by Brian Foote and Team Bulldog of Pacifica, California. Foote's creation stomped Robot X 41–4 in the first round of *BattleBots*' Season 3 lightweight tournament, and KO'd Cold Steel in Round 2 before being KO'd by Skid Mark in Round 3.

FAST FORWARD

With an emphasis on speed, Fast Forward was constructed by Phil Houdek of Mountain View, California. In the 2001 *BattleBots* competition in San Francisco, Fast Forward won a 32-13 decision over Annihilator, then lost a 23–22 squeaker to Twin Paradox.

FINTASTIC

Designed to resemble a hammerhead shark, this bot is a beauty, sleek and powerful. It was built by Edward Robinson, age thirty-nine, of Las Vegas, Nevada—who is a sword swallower, fire-eater, and heavy-equipment operator in his other lives.

Robinson has been messin' in bots for a couple of years and says that Robby the Robot is his all-time favorite Hollywood bot. Robinson's hobbies include renaissance fairs, rocketry, and hovercraft.

Fintastic made its public debut at *Robotica II* in 2001.

FIRE STORM

At the *Robot Wars* 3 (UK) tournament, Fire Storm won its opening heat with consecutive victories over Crasher, Knasher, Facet, and Diotoir. In the semifinal heat, Fire Storm continued to excel, bumping off Pit Bull and Panic Attack to make it into the final four. At that time it was defeated by the eventual tournament winner, Chaos II.

FIRE STORM 2

At the *Robot Wars* 4 (UK) tourney, Fire Storm 2 won its heat, defeating Bolt From the Blue and The Morgue.

FLAILBOT

Flailbot is a superheavyweight built by John Phillips of Avon, Colorado. In *BattleBots*' Season 3 tournament, Flailbot KO'd Ballbreaker in the opening round but was subsequently outpointed by Ogre 30–15.

FLATBOT

Talk about low riders, Flatbot is only four and a half inches high. Built in two months by Randal Boyd, Gabriel Sanchez Jr., and Paul Price of Gilbert, Arizona, Flatbot is built on wheels and is battery powered. A lightweight, Flatbot is a veteran of *BattleBots.* Its primary weapon is a saw blade.

FLEXY-FLYER

A superheavyweight at 210 pounds, Flexy-Flyer is a four-wheel-drive all-terrain robot, perfect for *Robotica,* where this bot was last seen. Flexy-Flyer has a prototype one-of-a-kind articulating chassis.

Flexy-Flyer has various wheel and tire setups, which are powered by four 1.6-horsepower motors using individual 20:1 reduction gearboxes. Flexy-Flyer has a lifting arm that is powered by a 1,000-pound force electric linear actuator.

During individual combat, Flexy-Flyer has a cow-catcher wedge attached to get under opponents. Its chassis, lifting arm, wedge, and bottom armor are made of aircraft-grade 4130 chrome moly and the covers are made of high-impact Lexan.

The bot was built by Mike Konshak, age fifty-five, of Louisville, Colorado, who also pilots it during combat. The weapon, the lifting arm, is operated by his teammate, Norm Domholt.

Konshak caught bot fever at the beginning of 2001 after watching robot combat programs on television, and by the end of the year, he was a veteran of one of those shows himself, competing at *BattleBots* and *Robotica II.*

In real life Mike is a senior advisory development engineer in client server disk engineering, and his hobbies include motorcycle racing, ham radio, and building a battery-powered home charged with photovoltaics. Right.

FLIP FLOP FLY

Say that three times fast. It ends up Floop Flap Fling. At the *Robot Wars* 3 (UK) competition, Flip Flop Fly defeated Wild Willy in the opening round of its heat before succumbing in the second round to Evil Weevil.

FLIPPER

The name Flipper did not come about by accident. Oh no. It was on porpoise. Flipper lost to Ultor in the opening round of the *Robot Wars* 3 (UK) tournament.

FLIP YOU OFF

"Flip You Over" would have been more accurate, but this name is so much more rude! Flip You Off was built by Ken Stevens and Team Squat Bot of Torrance, California. The bot received the bird rather than gave it in the opening round of the middleweight tournament at *BattleBots*' Season 3 TV taping, being KO'd by Little Drummer Boy.

FORKLIFT'S REVENGE

In the *Robot Wars* 3 (UK) tournament, Forklift's Revenge lost to Suicidal Tendencies in the opening round.

FORK-N-STEIN

Jim Snook of San Diego, California, is the fellow who screamed, "It's alive! IT'S ALIVE!" when Fork-N-Stein first came to life. After advancing in the first round past Dredge, Forky was knocked cold by Hexadecimator in Round 2.

FRENZY

It took more than 1,000 manhours to build Frenzy, a heavyweight veteran of the BattleBox. The main man who provided those manhours is Patrick Campbell of Stevenson Ranch, California. His teammates are Emil DeMillo, George Melero, and Joanna Campbell.

Frenzy is round, and its construction is mostly CNC-machined 6061-T6 aircraft aluminum. Using its titanium BattleAxe, Frenzy chops opponents down. That BattleAxe serves a dual purpose actually, as it can also be used to push the bot back onto its belly in case it gets flipped over.

In *BattleBots'* Season 2 heavyweight competition, Frenzy advanced as far as the semifinals, when it was eliminated by the eventual winner, BioHazard.

At *BattleBots'* Season 3 tournament, Frenzy won via a KO over Towering Inferno in the first round, then lost a 23–22 decison to OverKill in Round 2.

FROGHEMOTH

James Thomas of El Reno, Oklahoma, built Froghemoth, which was KO'd by Mantis in the opening round of *BattleBots'* Season 3 lightweight tournament.

FROSTBITE

Brrrrr, anyone who has seen the movie *Snow Day* knows just how menacing a snowplow in the wrong hands can be. Well here's a snowplow that is looking for trouble, and that is bad news for all of its opponents. Frost-

Bite, which took five weeks to build, calls its primary weapon its "snowplow of death."

Main builder Michael "Fuzzy" Mauldin of Irwin, Pennsylvania, says, "FrostBite features an invertible design with a hinge-mounted front snowplow blade and tires extending above the body that allow it to run equally well right-side up and upside down. It has four high-torque electric motors."

Mauldin built FrostBite with the help of Debby Mauldin, Danny Mauldin, Jacey Ross, and Kelsey Ross, otherwise known as Team Toad Lab. At the *BattleBots*' Season 2 Heavyweight competition, FrostBite knocked off the heavily favored OverKill in an early round.

There was irony in this victory, as OverKill was a Christian Carlberg bot and Mauldin learned how to build bots by looking at Carlberg's Web site.

That season, FrostBite was eventually bumped off by Voltronic. FrostBite KO'd Circuit Breaker in the opening round of *BattleBots*' Season 3 tournament, then lost a 23–22 decision to Hexadecimator in Round 2.

Frostbite has evolved and is currently in its third incarnation. Whereas the original bot has a fixed snowplow as its only, and not that effective, weapon, that was upgraded so that the plow blade now can be electrically actuated to attack from various angles.

The current bot is a "thresher" with a 1¼-inch solid steel shaft and nine solid steel teeth, each tooth weighing 4.6 pounds. Those are mounted with ANSI standard ¼-inch keyways held in place by split shaft collars.

FrostBite features a 9-horsepower MagMotor running at 36 volts, powered by 168 nickel metal hydride 3000 mAhr battery cells, all of which run the thresher. Those batteries can sustain over 300 amps for the three minutes a bot battle might take.

The power is transmitted from the motor to the weapon shaft using a size B V-belt, with a 2:1 ratio speed reduction. This gives the 55-pound rotating mass a speed of 2,800 revolutions per minute.

Custom-built wheelchair motors from National Power Chair give FrostBite its power. Bicycle chains connect the front and back wheels. Those chains give FrostBite the advantages of four-wheel drive without the necessity for four motors.

FROZEN TOAST

It was chilly in the BattleBox when Frozen Toast squared off against SnowFlake in the opening round of *BattleBots*' Season 3 lightweight tournament, but heated up quickly as SnowFlake quickly melted Frozen Toast, a bot built by Russell Burke and Team Evil Duck of Baltimore, Maryland.

FUSION

Constructed over a seven-month period by John Patrick, Thomas Gralewicz, and Team Saber of Crystal Lake, Illinois, Fusion was KO'd in the opening round of *BattleBots*' Season 3 middleweight tournament by Hazard. Its primary weapon is a pneumatic ram. Fusion works equally well upside down.

G

GADGET

The first bot ever built by John Kittelsrud and Team Droid, Gadget ran in the 1997 and 1998 BotBash competitions in Phoenix, Arizona. Gadget was killed in 1998 because its batteries were not strapped down inside. Live and learn. Gadget later broke down for keeps when too much current to the motors ended up toasting them, melting the communicators onto the shafts. Kittelsrud said he knew something was wrong when Gadget began to emit an odor like that of "Band-Aids melting."

GAMMA RAPTOR

The latest lightweight member of Robert Pitzer's Team Raptor family, this tread-wheeled bot won its first four matches at *BattleBots*' Season 3 lightweight tournament. It defeated SnowFlake 34–11, then KO'd Serial Box Killer and Mouser Super Mecha-Catbot. In the semifinals, the 58-pound Gamma Raptor defeated Wedge of Doom by a score of 28–17. In the finals, it lost a controversial 32–13 decision to Dr. Inferno Jr. The crowd thought Gamma Raptor was worthy of the big nut, but Team Raptor had to settle for second place.

GAMMATRON

Mike Okerman of Huntington Beach, California, is the main builder of Gammatron, a heavyweight bot that uses a combination five-inch titanium hammer and a saw blade as its weapon. Mike's teammates are Curt

Nemeth, Carlee Okerman, and Mike Routhier. In *BattleBots*' Season 3 heavyweight tournament, Gammatron was knocked unconscious in the opening round by Little Sister.

GANG GREEN WEASEL

Built by Michael Springer and Team UnderFight of Rock Springs, Wyoming, Gang Green Weasel knocked out Deathwishbone in the opening round of *BattleBots*' Season 3 heavyweight tournament, and then was KO'd by Slam Job in Round 2.

GARM

Garm is the creation of Don Lariviere and "Kampfgruppe5" of Redondo Beach, California. After knocking out Shovit in the opening round of *BattleBots*' Season 3 middleweight tournament, Garm was eliminated via a KO by Bad Attitude.

GARY GIZZMO

Mike Flanagan of Silvis Heights, Illinois, created Gary Gizzmo, which won a 32–13 decision over Velocity 1 in *BattleBots*' Season 3 middleweight tournament, then lost a 33–12 decision to Pack Raptors in Round 2.

GEMINI

At the *Robot Wars* 4 (UK) tourney, Gemini came into the competition seeded eighth. It defeated The Creature in the opening round of its heat but was defeated in the heat finals by Tornado.

GENERAL CARNAGE

General Carnage defeated Robo Cow in the opening round of the *Robot Wars* 3 (UK) tournament before losing to B.E.H.emoth in Round 2.

GENERAL CHOMPSALOT

Built by Tim Berghoefer, Cheryl Berghoefer, and Dave Grubszynski of New Berlin, Wisconsin, General Chompsalot weighs 216 pounds and measures forty-three inches by thirty-three inches by fifteen inches. It runs on two 36-volt electric motors, and operates equally well with either side up.

The front is a massive set of jaws that can bite and clamp down when open. When the mouth is closed the "upper lip" is a pointy poker. When the jaws open a big red tongue can be seen.

General Chompsalot competed in the Tag Team Terror Tournament during Season 1 of *Robot Wars: Extreme Warriors.* Teaming with Run Away, they defeated the team of Joker and Falcon in their opening match, then lost to Tut Tut and Drillzilla in the finals.

In the "American Revolution: War of Independence" episode of that same series, in which bots from the United States were pitted against bots from the United Kingdom, General Chompsalot went up against England's Stinger in the opening round. Stinger tried to use its spike to give Chompsalot a tonsilectomy and managed to knock off one of Chompsalot's eyeball hubcaps, but America was given the decision after it grabbed a mouthful of Stinger's tire and wouldn't let go. In the semifinals, an all-American matchup, Chompsalot took on the spinner named Joker. Chompsalot won a decision. In the opening contact of the bout, Chompsalot managed to bite off one of Joker's mace weapons and the opponent ended up in "The Zone," being eaten by the house bots. That sent Chompsalot to the finals, where it was defeated by the British Pussycat. Pussycat's blade chewed up Chompsalot real good, as they say, while Chompsalot could only toothlessly gum Pussycat.

GENERAL GAU

Built by William Garcia and Team Goosebearys of Boston, Massachusetts, General Gau was demoted via KO by Black Warrior in the opening round of *BattleBots*' Season 3 lightweight tournament.

GINSU

Ginsu was the selected victim for an exhibition bout held against Chin-Killa, the bot sponsored by and designed after the face of *Tonight Show* host Jay Leno. Ginsu, in this match, was flipped onto its back by the "Power of the Chin."

GOATHAMMER

A gruesome hybrid of animal and weapon, Goathammer is the unholy creation of Timothy Ray and the Goat Brothers of Oak Park, California. In the opening round of *BattleBots*' Season 3 middleweight tournament, Goathammer was outpointed and eliminated by Traxx, 27–18.

GODTILLA

Maybe in Japan this bot might have been able to cause some mass destruction, but it was relatively harmless in the BattleBox, being knocked out in the opening round of *BattleBots'* Season 3 lightweight tournament by Daisy. Built by Anthony Speros and Lunchbreak of York, Pennsylvania.

GOLDDIGGER

Hi-ho, hi-ho, it's off to the BattleBox we go. But look out for Golddigger's weapon, a nasty pickax, that can bust up a robot just as easily as rock in a mine. Built in one month by Randy Eubanks and David Ray of Long Beach, California, Golddigger fights in the heavyweight division.

At *BattleBots'* Season 2 heavyweight contest, Golddigger won its opening match against KillerHurtz when it used its weapon to whack its opponent all around the BattleBox, but was eliminated in the second round. In *BattleBots'* Season 3 tournament, Golddigger won a 30–15 decision over Scrap Daddy's Persistor in the opening round, defeated Kritical Mass II 23–22 in Round 2, and then was defeated soundly (38–7) by Tazbot in Round 3.

GO POSTAL

Who better to build a bot named Go Postal than a lunatic named Bernardo Luna, he of the San Diego, California, Lunas? Go Postal is a superheavyweight that lost in the first round of *BattleBots'* Season 3 tournament, outpointed by Phere 25–20 .

G.O.R.T.

Named after the awesome robot from *The Day the Earth Stood Still,* this G.O.R.T. is the creation of Kelly Osborn of Fort Worth, Texas. The name is cooler than the bot, which competed ineffectively in *BattleBots'* Season 3 tournament.

GRAVEDIGGER

At the *Robot Wars* 3 (UK) competition, Gravedigger won its opening heat, defeating Manic Mutant, Mortis, and Darke Destroyer in the process. In the semifinal heat, Gravedigger lost in the opening round to Steg-a-Saw-Us.

GRAY MATTER

Israel Mathewson of Medford, Oregon, really used his brains when building this robot. Maybe that's why he called it Gray Matter, which was built, says Mathewson, over "four months of Saturdays." Team Gray Matter consists of Randy Mathewson, Noah Mathewson, and Sally Mathewson.

The bot's main weapon is a sixteen-inch spike mounted to the front bumper. A superheavyweight, Gray Matter uses battering-ram techniques to take out its opponents. It has what Israel calls "high mass hitting power."

At *BattleBots'* Season 1 superheavyweight tournament, Gray Matter was eliminated in the opening round by the eventual winner, Minion. In the Season 2 tourney, Gray Matter was eliminated in a shoving match against Revision Z in the opening round. In the Season 3 tournament, Gray Matter lost in the first round by KO to Revision Z.

GREEN DRAGON

John Pagano, Deborah Pagano, and Team Dragon Combat Robots of Mt. Laurel, New Jersey, built Green Dragon, which was knocked out by Death by Monkeys in the opening round of *BattleBots'* Season 3 lightweight tournament.

Green Dragon features bulletproof Lexan armor. It operates equally well regardless of which side is up. Its primary weapon is a spinning saw blade in the front. Green Dragon took six months to build.

THE GREEN HOUSE

Built by Jim Moulton and David and Kristi Simmons of Tyler, Texas, The Green House is a highly maneuverable 180-pounder, which operates on two electric motors and features a pneumatic lifter as its main weapon.

The Green House's weakness has turned out to be the fact that it cannot self-right. Flip it onto its back and, as the Crocodile Hunter would say, "It's all ovah!" In the "Robot Rebellion" episode of *Robot Wars: Extreme Warriors,* in which entrants competed for the right to take on the fearsome *Robot Wars'* house bots, The Green House was defeated—shoved into "the pit," in fact—in the eliminator round by Tut Tut.

THE GREEN MOUSE

Another competitor in the "Robot Rebellion" episode of *Robot Wars: Extreme Warriors,* The Green Mouse is covered with green fur and is flammable. If there is one cardinal rule on *Robot Wars,* it's that things that can burn always do. Why else have a house bot that breathes fire like Godzilla?

The Green Mouse was built by Evan White, Terry White, and Jim Mover of Orange County, California. It weighs 198 pounds, and measures fifty-five inches by twenty-two inches by sixteen inches. It is powered by a pair of electric motors and has a proven-effective pneumatic flipper as its primary weapon.

At the "Robot Rebellion," however, The Green Mouse was the flippee rather than the flipper. Losing in the eliminator round, The Green Mouse was defeated by Tricerabot. The emerald rodent was thrown onto its back and then set afire.

GREENSPAN

Greenspan, constructed by Steven Brown of Anaheim, California, defeated Administrator in the opening round of *BattleBots*' Season 3 heavyweight tournament by a score of 27–18. The bot then knocked out I-Beam in Round 2, but lost by KO to Voltronic in the third round.

GRENDEL

Grendel's main builder is Daniel Rupert. He was helped by Scott Robeson, Steven Schulze, and Seth Herring, all of San Diego, California. A participant in *BattleBots*' Season 1 superheavyweight tournament, Grendel was defeated in the second round by the eventual winner, Minion. Grendel was revamped and made future *BattleBots* appearances under the name Revision Z. (See also REVISION Z.)

GRIFFON

At the *Robot Wars* 3 (UK) competition, Griffon defeated Daisy Chopper in the opening round of its heat, then lost in the second round to Cerberus.

GRIMLOCK

Software engineers Andre C. Anderson, age twenty-seven, and Marshal Hagen, twenty-six, both of Seattle, Washington, have been tinkering around with robotics for ten years, the last three of which have been hardcore.

Their robot is called Grimlock, a 200-pound robot with a two-wheel differential drive base. It is powered by a pair of 24-volt wheelchair motors with two 4QD 100-amp speed controllers.

Grimlock's weapon is a pair of powerful hydraulic jaws that can wrap themselves around an opponent and chomp down until the hapless foe has been crushed flat like a spatula. Both the chassis and the jaws are made of salvaged steel.

In addition to *BattleBots,* Grimlock has battled at the 1998 Robot Wars, the 1999 Seattle Robotics Society/LEGO Robot Wars, where Grimlock earned a second place, and the 1999 Seattle Robotics Society Sumo 3kg Class (fourth place). In *Robotica*'s Season 1 tournament, Grimlock lost in the preliminary rounds to Kritical Mass.

GRIM REAPER

At the *Robot Wars* 3 (UK) tournament, Grim Reaper was eliminated in the opening round of its heat by the eventual heat winner, Big Brother.

GUNGNIRE

Michael Bastoni and the Pilgrim Nuclear Technology Alliance of Plymouth, Massachusetts, is responsible for Gungnire, which won a 23–22 decision over Tower of Power in the first round of *BattleBots*' Season 3 middleweight tournament, then was KO'd in Round 2 by Evil Con Carne.

GYREX

Built by Nicolas Peavy of Texarkana, Arkansas, this superheavyweight lost in the first round of *BattleBots*' Season 3 tournament. Gyrex was defeated in a close decision (24–21) to Dawn of Destruction.

H

HALF ASS AND RANDOM

With a name like that, it is easy to underestimate this bot, which was built by David Robison and Team Measure Once Cut Twice of Buena Park, California. But Half Ass and Random scored a KO victory over Wham! Bam! in the opening round of *BattleBots*' Season 3 lightweight tournament before losing via KO in the second round to Skid Mark.

HALO

Brian Scearce and Team Second Breakdown of Fremont, California, created Halo, which was smashed real good by T-Minus in the first round of *BattleBots*' Season 3 middleweight tournament.

HAMERSCHLAG

Hamerschlag was built by a pair of nineteen-year-old Carnegie Mellon University students, Dan Haeg of Minneapolis, Minnesota, and Zach Menegakis of Boca Raton, Florida.

The robot's chassis is made up of six 4-inch-by-4-inch-by-⅛-inch square aluminum tubes with a ⅜-inch polycarbonate cover. The power plant consists of four DeWalt nicad drill batteries and four DeWalt cordless drill motors connected to six 5-inch wheels with a ½-inch-thick cogged V-belt screwed in. They are controlled by two Innovation First Victor speed controllers and a Futaba 8U Super radio.

Hamerschlag is a *BattleBots* veteran.

HAMMERHEAD

As the name suggests, there is nothing subtle about HammerHead's approach to robotic competition. It runs into its opponent, then it backs up and runs into it again. Its primary weapon is ram power.

A lightweight bot built over a two-month period by Charles Steinkuehler and Team Brute Force of Topeka, Kansas, HammerHead knocked out Rim Tin Tin in the opening round of *BattleBots*' Season 3 lightweight tournament before losing a 45–0 shutout to Wedge of Doom in Round 2.

At the *Robot Wars* 3 (UK) competition, HammerHead lost in the opening round of its heat to Pussycat.

Steinkuehler describes the bot's strong points: "High-strength frame. Powerful high-torque/low-inertia industrial servo motors. Massive amounts of niCad battery power. Nearly indestructible . . . nothing but batteries, motors and aluminum."

HAMMERLOCK

Created by Tom Wickey of Galveston, Texas, this superheavyweight lost in the first round of the *BattleBots*' Season 3 tournament to Whyachi.

HAMMERTIME

Team Hammertime is led by Jerry Clarkin of Coatesville, Pennsylvania. In *BattleBots'* Season 3 lightweight tournament, Hammertime, which weighs only 20 pounds, did very well. Its primary weapon is a pneumatic spike. After defeating Maximus in the opening round by a 31–14 decision, Hammertime KO'd Brainsters of Triskellion in the second round before losing a 30–15 decision to Vladiator in Round 3.

HAMMERTRON

Lost in the opening round of the *Robot Wars* 3 (UK) tournament to Sonic.

HAMUNAPTRA

Built over a month-and-a-half span by Nathan Poort and Dominic Vannucci of Las Vegas, Nevada, Hamunaptra uses a swinging hammer to pound its foes. Fighting in the superheavyweight division, Hamunaptra also features skid steering, an interchangeable weapon design, and a low-gravity design.

HANDSOFF!

HandsOff!, a veteran of the BattleBots 2000 and DragonCon 2001 competitions, was built by brothers Greg and Chris Hannold, their father, Richard Hannold, and Greg's wife, Janella, all from Salisbury, North Carolina. The robot features four 24-volt motors, 16:1 gear ratio, and ten-inch go-kart slicks.

Its primary weapon is a nine-inch stainless steel spike known as the Rotor of Death. The bot is armored with a half inch of Lexan. The radio-control mechanism is comprised of a Hitec seven-channel PCM remote, and 4QD Pro-120 speed controllers.

In *Robotica*'s Season 1 tournament, HandsOff! was defeated in the preliminary rounds by Ram Force.

HAZARD

Hazard was built by Tony Buchignani of Los Angeles, California. Buchignani is a lawyer during the day.

Hazard has a forty-inch hardened steel propeller blade as its primary weapon. It won the big nut in *BattleBots'* Season 1 middleweight tournament, defeating Turtle Road Kill, Spin Orbiting Force and, in the finals, Deadblow, along the way.

In *BattleBots'* Season 3 tournament, it successfully defended its championship. Hazard KO'd Fusion, scored a 34–11 victory over Zion, then eliminated T-Wrex and Little Drummer Boy to win the Nut.

More about Team Hazard can be found at www.legalword.com.

HEADTRIMMER

Paul Bell and Team Springcrest of Chapel Hill, North Carolina, built Headtrimmer, which lost by KO to Snap Dragon in the opening round of *BattleBots'* Season 3 lightweight tournament.

THE HEART OF DARKNESS

Ever been to Saigon? Well, conjuring up images of surfing to The Doors and napalm in the morning, here's a bot that will make you feel as if you have. The Heart of Darkness was built by Patrick McCarthy and Team Apocalypse Now of Reston, Virginia. In *BattleBots'* Season 3 heavyweight tournament, The Heart of Darkness KO'd Jack Tripper in the first round, then was KO'd in turn by Shark Byte in Round 2.

HEFTY

At the *Robot Wars* 3 (UK) tournament, Hefty lost in the opening round of its heat to X Terminator.

HENRY

Don't call him Hank! Hammerin' Henry defeated Haardvark in the opening round of its heat at the *Robot Wars* 3 (UK) competition before losing in the second round to Steg-a-Saw-Us.

HENRY II

At the *Robot Wars* 4 (UK) tourney, Henry II was eliminated in the opening round of its heat by the eventual heat winner, Dominator 2.

HERR GEPOÜNDEN

One of the finest lightweights at *BattleBots* Season 3. Built by David Otto and Team Zwölfpack of Santa Ana, California, Herr Gepoünden won five consecutive bouts before being eliminated by Dr. Inferno Jr. in a 27–18 decision. The five wins were over, in order, Centipede (26–19), Doomsday Machine (36–9), Scrap Daddy's Surplus (30–15), The Crusher (KO), and Death by Monkeys (KO).

HEXADECIMATOR

Built by Tim Paterson and Team WhoopAss of Redmond, Washington, Hexadecimator made a spectacular showing at *BattleBots*' Season 3 heavyweight tournament. It won five bouts, three of them by KO, before being eliminated.

In the opening round, Hexadecimator KO'd That Good R&D. It KO'd Fork-N-Stein in Round 2, and won a 23–22 decision over FrostBite. It KO'd Shark Byte and outpointed KillerHurtz by a score of 32–13, then lost by a KO to Son of Whyachi.

HIPOPONONOMUS

You can stare at it and stare at it and you still won't be able to pronounce it. Hipopononomus, which is unfortunately pronounced just the way it looks, is the creation of Justin Chou and Team Batteraizees.

The bot might be unpronouncable, but the polysyllabic Hipopononomus was not undefeatable. In fact it was KO'd in the opening round of *BattleBots*' Season 3 middleweight tournament by Zion.

HOBGOBLIN LTD.

Built by Chad Campbell and Team Warlock of Huntley, Montana, Hobgoblin Ltd. lost in the opening round of *BattleBots*' Season 3 lightweight tourney, 28–17, to Wedge-o-Matic.

HOOBOT

Hoobot made it to the third round of *BattleBots*' Season 3 superheavyweight tournament. After a razor-thin points victory over Steel Reign in the first round (23–22), Hoobot defeated Trimangle 27–18 before losing to Ogre by a score of 34–11 in the third round. Hoobot was built by Rocky Mohacsi of San Mateo, California.

HORRIFICA

Brian Patrick and Team Beta of Enumclaw, Washington, built Horrifica, which outpointed Kegerator 27–18 in the opening round of *BattleBots'* Season 3 middleweight tournament. In the second round it KO'd SMD, then lost via KO to Eraser in Round 3.

HOT WHEELS

Hot Wheels' offense is threefold. First of all, it is a wedge and can get under its opponents. It also has a hinged front door that can both lift an opponent or cut him with the door's sharp edges. And, if all else fails, it has a top-mounted accessory that can launch a nine-foot-square net onto its foe.

Hot Wheels, as its name suggests, is a speedy bot, traveling up to sixteen miles per hour. One-quarter-inch aluminum plate was used to fabricate the chassis. Builders Mark Mayer and Eric Ostendoff—both forty-year-olds from Torrence, California, who work as toy designers for Mattel Toys in Los Angeles—have used .090 aluminum to build the body shell and have installed a Lexan window on the top. Hot Wheels uses 6- and 12-volt gel cells with ESCs to control five motors for drive and steering.

Eric got his start in robotics back in 1994 when he won first place in a Trinity College contest in Connecticut. In the contest, an autonomous robot had to locate a candle in a maze and extinguish it.

In *Robotica*'s Season 1 tournament, Hot Wheels made it past the preliminary rounds but was eliminated in the Fight to the Finish by Kritical Mass.

HUGGY BEAR

Not all bots have fearsome names. This bot proves that you can't judge a machine by its monicker. Huggy Bear scored three victories in *BattleBots'* Season 3 middleweight tournament. It KO'd Portable Killsaw in the opening round, won a 32–13 decision over Iron Soldier in Round 2, then KO'd Timmy before losing a 23–22 squeaker in the fourth round to Sabotage.

HYPNODISC

Hypnodisc's favorite singing group? The Spinners, of course. Hypnodisc won its heat at the *Robot Wars* 3 (UK) competition with consecutive victories over Robogedden, Stealth, and Beserk II. Hypnodisc also emerged victorious in the semifinal heat of eight, defeating Evil Weevil, 101, and

Steg-a-Saw-Us to get into the final match. Hypnodisc had to settle for second place in the tournament, however, as it was vanquished in the finals by Chaos II.

Hypnodisc is the baby of Dave Rose. The bot's spinning blade has proven tough to beat. Dave first became interested in battling robots when he watched *Robot Wars* Series 2 in England.

Hypnodisc's spinning blade has grown steadily more menacing as the competitions have passed. In *Robot Wars* Series 3, Hypnodisc's disc went round at 500 revolutions per minute; in Series 4 it was 750 revolutions per minute; for Series 5 the blade spun at 900 revolutions per minute.

But there is a danger in making the spinning weapon too fast. As Dave explains, "The speed of the disc is now at the point that a really good contact means Hypnodisc ricochets violently, putting large stresses on every part of our robot."

Dave gets all of his welding done by his friend Mark, who once worked for NASA welding parts for the space shuttle.

I

I-BEAM

Built by Dean Du Bois and Team Girder of Clearwater, Florida, I-Beam knocked out Chomp in the opening round of *BattleBots*' Season 3 heavyweight tournament, then lost by KO in Round 2 to Greenspan.

ICEBERG

Who better to commit cold-blooded roboticide than a killer named IceBerg? In *BattleBots*' Season 3 superheavyweight tournament, however, IceBerg merely froze up and lost in the opening round to The Judge by a whopping margin of 38–7.

ICE CUBE

Debbie Mauldin and Team Crystal of Irwin, Pennsylvania, built Ice Cube, which scored a decisive 40–5 win over Pink Slip in the opening round of *BattleBots*' Season 3 middleweight tournament. It then defeated Die Fledermaus by a score of 33–12 in Round 2 before being eliminated by Little Drummer Boy in a 29–16 decision in Round 3.

ICKY THUMP

Ewwwww, Icky Thump! I'm not sure what the name means, but it gives me the willies! Icky Thump was begat by Mike Olsen of Davenport, Iowa. In *BattleBots*' Season 3 heavyweight tournament, Icky Thump thwacked The Ripper, winning a 24–21 decision. Icky Thump was subsequently thwarted by Junkyard Offspring, losing a 25–20 decision in Round 2.

I-DROID

Walter Martinez, a systems and network administrator at California State University at Long Beach, built the 210-pound I-droid. I-droid's weapons consist of a stainless steel pitchfork—to which additional spikes have been added—for piercing and deep piercing, a.k.a. impaling. Two elephant-like tusks protrude from the front of the bot. They move up and down and can lift and sometimes flip an opponent.

The menu of items that make up I-droid includes two 24-volt gel cells, one 6-volt SLA, Vantec ESC, two 24-volt electric wheelchair motors, a 12-volt linear actuator, a 12-volt electric wheelchair motor, two Team Delta (TD) R/C H-bridges, one TD Switch, a six-channel RC RX/TX, aluminum, steel, and six wheels. Along with *BattleBots,* I-droid has also competed in the Westec '96 Robotics Challenge.

ILL-TEMPERED MUTT

Weighing in at 210 pounds, this cranky K9 operates on two-wheel drive and is invertible. It has been timed at twenty-six miles per hour, which is pretty doggone zippy, and it changes directions well also.

At *Robotica II,* Ill-Tempered Mutt operated as a wedge for the preliminary rounds, but, after getting to the Fight to the Finish, changed configurations and went to war with a full set of teeth, composed of a thirty-pound steel drum with beater bars on it spinning backward at almost 2,500 revolutions per minute.

The Mutt was built by Ken Benner, age thirty-seven, of Lake Mathews, California. Ken is an electrician and is studying to become a science teacher. His hobbies include high-voltage experiments and flying small planes. His teammate is Jim Evans.

INCISOR

Incisor's favorite game show? *To Tell the Tooth,* of course. Incisor was constructed by Kevin McEnery and Pain USA of Fox Lake, Illinois. In *BattleBots*' Season 3 heavyweight tournament, Incisor put the bite to Maximum Overdrive in the opening round and won by a KO. In the second round Incisor was defeated by Center Punch, losing by a score of 29–16.

INCOMING

Here is a heavyweight bot to be reckoned with. Incoming was constructed by Brian Moore and Anger Management, Inc., of Cambridge, Ontario, Canada. In the opening round of *BattleBots*' Season 3 heavyweight tournament, Incoming KO'd Sgt. Spanky. It followed that victory up with another KO win over Parasitic in Round 2. Incoming won a 25–20 decision over Mauler 5150 in the third round and then was KO'd by BioHazard in Round 4.

INDEFATIGABLE

At the *Robot Wars* 4 (UK) tourney, Indefatigable lost in the opening round of its heat to Atomic.

INSTIGATOR

Dennis Millard, Scott Fullen, and Team Byte Me of Salt Lake City, Utah, built Instigator, which was KO'd in the opening round of *BattleBots*' Season 3 middleweight tournament by Blade Runner.

Instigator took six months to build. It is flippable and rests on its wheels no matter which side is up. Its primary weapon is a horizontal kinetic energy disc.

INTRIVERTER

InTriVerter is a three-wheeled invertible bot that rolls about on 16-inch tires on steel rims. It is powered by two E-Tek motors at 36 volts delivering over 20 horsepower combined.

The bot can pick 'em up and put 'em down, registering on the speedometer at faster than thirty miles per hour. It is armored with ⅜-inch aluminum and ⅛-inch titanium.

InTriVerter was built by lawn-mower mechanic Ted Shimoda, age thirty-seven, and Richard Blom, age thirty-four, of Pasadena, California. This pair have been working in robotics for ten years. Shimoda's hobbies include ocean fishing, snowboarding, sailing, and jet skiing, as well as radio-controlled cars and planes. Also on the team is Kevin Smith.

INVERTIBRAT

Invertibrat, with the emphasis on the *brat,* defeated The Witch in the opening round of its heat at the *Robot Wars* 3 (UK) competition, then whupped Terrapin in the second round. In the heat finals, this bot was eliminated by Beast of Bodmin. At the *Robot Wars* 4 (UK) tourney, it was defeated in the opening round of its heat by The Creature.

IRON AWE

At the *Robot Wars* 4 (UK) tourney, Iron Awe defeated Mazakari in the opening round of its heat, then lost to the eventual heat winner, Steg 2, in the second round.

IRON EAGLE

All the way from Kawasaki City, Japan, comes Iron Eagle, a wheeled and battery-powered bot. It took main builder Terukazu Nishimura two months to build the Eagle, with help from teammate Takayoshi Vemitsu. The bot's primary weapon is the simulated beak of an eagle.

THE IRON MASK

At the *Robot Wars* 3 (UK) competition, The Iron Mask lost in the opening round of its heat to Weld-Dor.

IRON SOLDIER

West Springfield, Massachusetts's, Jason Demerski and Team ACMX built Iron Soldier, which marched to a 41–4 decision over Malvolio in the opening round of *BattleBots*' Season 3 middleweight tournament. In Round 2, however, Iron Soldier was killed in action, losing a 32–13 decision to the fearsome Huggy Bear.

ISOSCELES 1

Built by Eric Koss of Virginia Beach, Virginia, Isosceles 1 was KO'd in the opening round of *BattleBots*' Season 3 superheavyweight tournament. Isosceles 1 was rendered motionless by No Apologies.

I-WILL-SMACK-U

Robert MacDonald and the team he calls Shaddup of Bethlehem, Pennsylvania, built I-Will-Smack-U, which won a pair of knockout victories at *BattleBots*' Season 3 lightweight tourney before losing via KO to Bad Habit.

J

JACK THE GRIPPER

Jack the Gripper does not prowl the night streets of London, but rather those of Downey, California, where it was constructed by Anthony King. A superheavyweight, Jack the Gripper was eliminated in the opening round of *BattleBots*' Season 3 tournament when it was KO'd by Cuad the Crusher.

THE JACK TRIPPER

David Robison and a team he calls Measure Once, Cut Twice of Buena Park, California, built The Jack Tripper, which was KO'd by The Heart of Darkness in the opening round of *BattleBots*' Season 3 heavyweight tournament.

JAV MAN

With a bunch of these and General George Patton you could march across Europe. Jav Man is engineered like a real battle tank. Jav Man has rubber-cover titanium treads and a 60-volt industrial drive system.

The drivetrain and frame are completely custom fabricated from 7075, 6061, and 7068 aircraft aluminum and 6al4v titanium. There are no welds used in its construction, with the exception of the exterior bumper guards.

Jav Man was built by Greg Ferree, age forty-nine, of Conneaut Lake, Pennsylvania, who is a self-employed industrial electronics service technician with a side interest in sailboat racing. Greg says that his favorite robots are the theme-park robots such as those at Disney, Busch Gardens, etc. His teammates are Dee Ferree and Tim McKain.

Jav Man made its competitive debut at *Robotica II*.

JAWBREAKER

A heavyweight at 210 pounds, Jawbreaker is built so that it looks as if it has a face. And, just like the name implies, you had better look out for the mouth. Jawbreaker has only two wheels but can go a maximum of ten miles per hour and can develop a half ton of force in its jaws. When the mouth is all the way open, robots up to thirty inches tall can be scooped up inside it. Jawbreaker's jaws and frame are made from steel tubing, the armor from extruded aluminum, and the teeth are titanium.

Jawbreaker can function upside down, but since this renders all of the radio controls backward, it tends to cut down on its mobility. He is the creation of thirty-one-year-old Tim Berghoefer of New Berlin, Wisconsin, a machine-repair journeyman who collects pinball machines.

"My bot is built like a Sherman tank," boasts Berghoefer.

In *Robotica*'s Season 1 tournament, Jawbreaker defeated Crisis Management in the preliminary rounds, then took out Windingo in the Fight to the Finish. This earned Jawbreaker a place in the Final Six, where it took on Ram Force in the preliminary rounds. Jawbreaker won the Gauntlet, but lost on both the Speedway and the Maze to be eliminated by a score of 250–160.

JAWS OF DEATH

The raw chomp power of Jaws of Death was strong enough to get this superheavyweight into the second round of *BattleBots*' Season 3 tournament. After defeating KillDozer 29–16, Jaws of Death—built by Curt Meyers of San Diego, California—was defeated by Electric Lunch 30–15.

JOKER

The creation of Skip Harmeyer, Dane Harmeyer, and Joe Schermoerhorn—The Sylvan Lake Robot Warriors—from Rome City, Indiana, Joker weighs 215 pounds and measures thirty-two inches by thirty-two inches by nineteen inches. It runs on two electric motors. A round spinner, its weapons are maces attached to its outside that extend and dish out blows when the bot spins. One weakness of Joker's is that, if toppled, it lacks the ability to get itself right-side up again.

Joker competed in the Tag Team Terror Tournament during *Robot Wars: Extreme Warriors* premiere season. Its teammate was Falcon. Together the team lost its opening match to General Chompsalot and Run Away, and then, in the battle of losers, it lost again, this time to Rosie the Riveter and Tricerabot.

In the "American Revolution: War of Independence" episode of *Robot Wars: Extreme Warriors,* in which bots from the United States were pitted against bots from the United Kingdom, Joker KO'd Spawn Again in the opening round. One smack from Joker and Spawn Again broke down. In the semifinals, an all-American contest, Joker had one of its mace weapons bitten off by the hungry jaws of General Chompsalot in the opening moments of the bout and never recovered, losing a decision.

JOLLY ROGER

You knew from the start that the ol' skull and crossbones was going to be a heavily used design element of fighting robots—and nowhere more so than here. This pirate bot was built by John Olson and Team Tenacity of Memphis, Tennessee, and lost by KO to RoboSapien in *BattleBots*' Season 3 heavyweight tournament.

JO MAMA

Norman Muzzy and Team Flaming Monkeys of Cedar Falls, Iowa, possess the feverish minds that created Jo Mama. And its back to the drawing board for them, as Jo Mama forfeited her opening match in *BattleBots*' Season 3 lightweight tourney to Blood Dragon.

THE JUDGE

An exceptional superheavyweight built by Jascha Little of Austin, Texas, The Judge made it to the fourth round of *BattleBots*' Season 3 tournament. It defeated IceBerg 38–7, and knocked out Wack-a-tron and No Apologies before losing by KO to Rammstein in Round 4. The Judge won the Best Engineered Robot and Most Aggressive Robot awards at the tournament.

JUDGE SHRED

At the *Robot Wars* 3 (UK) tournament, Judge Shred wielded a mean gavel indeed in the opening round of its heat, defeating Mr. Punch. His Honor was then eliminated in the heat's second round by X Terminator.

JUGGERBOT

JuggerBot's 210 pounds are almost all comprised of its drive propulsion and structural reinforcement. Builders include aviation buff Mike Morrow, age thirty-one, Chris Gattman, age thirty-two, Tom Vaeretti, age thirty-three, and Ron Ender, age thirty-seven, of Corvalis and Portland, Oregon. They love to brag about JuggerBot's onboard thermal telemetry systems, which monitor the all-wheel drive and fully redundant radio systems. The bot has skid steering, which enables a zero turning radius. JuggerBot's powerful drive motors provide a thrust-to-weight ratio, it is claimed, that exceeds that of the space shuttle. It is a battering ram, built to have both speed and power.

What do the members of Team JuggerBot do for a living? Well, Mike Morrow is a manufacturing process engineer for Hewlett-Packard; Chris Gattman, along with being a musician, is a technical writer and the editor in chief of *JAM* magazine; Tom Vaeretti is a mechanical engineering student at Oregon State University; and Ron Ender is a design engineer for Hewlett-Packard.

At *BattleBots*' Season 3 superheavyweight tournament, JuggerBot knocked out Neptune in the opening round before losing a 28–17 decision to Vladiator.

At *Robotica*'s Season 1 tournament, it performed very well, beating Noll in the preliminary events, then taking out Hamerschlag in the Fight to the Finish. This earned JuggerBot a spot in *Robotica*'s Final Six. In that round, JuggerBot faced off against Viper in the preliminary rounds, and won all three events to earn a 250–125 points victory and a spot in the three-way final Fight to the Finish where JuggerBot was eliminated by the tournament winner, Run Amok.

THE JUNKYARD COG

Hey, there's nothing meaner than the Junkyard Cog. The creation of Jerame Powell of Phoenix, Arizona, The Junkyard Cog defeated Super Shark in the opening round of *BattleBots*' Season 3 superheavyweight tournament. In Round 2, the bot KO'd Africanized Killer Bot before being soundly outpointed 37–8 by World Peace in the third round.

JUNKYARD DOG

Team Dog House and head builder Kenneth Demshki of Torrance, California, built Junkyard Dog, which, appropriate to its name, scored a pair of KO victories in *BattleBots'* Season 3 lightweight tourney before it was eliminated in the third round by Serial Box Killer, coming up on the short end of a 24–21 decision. Dog's wins were over Dark Steel and DiamondBack.

JUNKYARD OFFSPRING

Obviously, a love child. Junkyard Offspring was fathered by Anthony Elder from Auburn, California, along with his team of Junkyard Dogs. At *BattleBots'* Season 3 heavyweight tournament, Junkyard Offspring KO'd Disco Cow 3000 in the opening round, then won a 25–20 decision over Icky Thump before losing by an identical score to Bacchus in Round 3.

K

KANUCKLE BUSTER

This bot, patterned after the movements of an ape, thus the name, is known for its agility. It has grip. That's because of its fifteen-inch custom-built wheels, which offer the bot incredible traction and excellent ground clearance.

Ground clearance is especially important when competing on *Robotica,* as Kanuckle Buster was last seen doing.

Kanuckle Buster's flipping plow turns into a battering ram to bust through the walls and forest of glass. Of course, that arm can also be used to flip the other robots. The bot also has arms that allow it to lift itself over obstructions.

Kanuckle Buster was built by Ron Woodward, age forty-six, Matt Woodward, age twenty-one, Stephen Rawls, age sixteen, Phillip Au, age nineteen, and Tom Fairchild, age eighteen—all proud residents of Yorktown, Virginia.

The gang has been at it for three years and, before hitting the big time at

Robotica II, they had gained valuable experience at the Trinity Fire Fighting Robot in 1998 and 1999, where they did not place; the First Robotics Mentor team 122, where they were the 2000 winner; the Richmond VCU Regional, and First Robotics Mentor team 122, 2001 Nationals, where they were division champ, and fourth in the nation.

KATER KILLER

At the *Robot Wars* 3 (UK) competition, Kater Killer lost in the opening round of its heat to Napalm. At the *Robot Wars* 4 (UK) tourney, it was eliminated in the opening round by the eventual heat winner, Tornado.

KEGERATOR

You know the deal: you rip out the insides of your old refrigerator, which has perhaps been moved into the basement, then punch a hole in the door. Inside goes the keg, outside goes the tap, and you're all set: Kegerator. Good for watching three straight football games—but for a bot, not so great.

Kegerator, which was built by Northrup David of Concord, New Hampshire, lost a 27–18 decision to Horrifica in the opening round of *BattleBots*' Season 3 middleweight tournament.

KEGGER

This middleweight veteran of *BattleBots* was built in two solid weeks of work by Nathan Roseborrough, Carl Lantz, and Jay Banningfield of Lubbock, Texas. Its weapons are a high-speed bashing arm and a stationary impaling spike. The bot has the ability to do the "vortex tornado spin of death," rotating an attached weight at high speeds.

KENNY'S REVENGE

Battling in the superheavyweight division is Kenny's Revenge, a wheeled and battery-powered bot that uses powerful saws as its main weapons. Main builder Thomas Kohlert of St. Charles, Illinois, lists the bot's other main feature as its low center of gravity. Kohlert's teammate is Karon Impson. Kenny's Revenge took two months to build.

KILL-9

Bob Proctor and Chaos Bots of Ottawa, Ontario, Canada, created Kill-9, which lost a 32–13 decision in the opening round of *BattleBots*' Season 3 middleweight tournament.

THE KILLA GORILLA

Magilla's nasty brother. Last seen competing in *Robotica II,* this bot is designed specifically for that show's revamped challenges. In order to get past obstacles, it has wheels that are thirteen inches in diameter and offer a full three inches of ground clearance.

A speedy primate, The Killa Gorilla has been clocked at twenty miles per hour. Its lifting arm not only has the muscle to lift more than 300 pounds, but, when it fully extends, it can be used to get the bot back on its wheels if it gets flipped over.

The Death by Monkeys team is headed by team captain Robert Farrow, who is a behavioral researcher. Also on the team is Pam Klein Farrow, a graduate student studying primate development, Steve Judd, who is retired, and Nora Judd, an artist. They are from either Seattle or Redmond, Washington.

KILLDOZER

Team KillDozer is headed by Michael Brouillard of Bay Point, California. A superheavyweight, KillDozer lost in Round 1 of *BattleBots*' Season 3 tournament to Jaws of Death. The bot was outpointed 29–16.

KILLERB

Check this out: top speed for KillerB, forty-five miles per hour! A wedge-shaped bot that is so fast and agile that it has defeated more powerful opponents, KillerB is also unusual because it is bimodal. That means that it can operate in two configurations. The robot has a tubular steel frame and aluminum armor plate.

KillerB was built by computer programmer and engineer Chuck McManis, age forty, Ellen McManus, age ten, Jim Gadd from Sunnyvale, California, "Steamboat" Ed Haas, age fifty-four, of Santa Rosa, California, and Mark Kottlowski of San Jose, California.

KillerB has competed in many shows, including the Robotics Society of America's Robot Games in 1994 and 1995, the San Jose State Sumo

Contest of 1993, the Austin Robofest 5 in 1994, and BEAM Robotics in Santa Fe, New Mexico, in 1995.

At *BattleBots*' Season 3 heavyweight tournament, KillerB won a lopsided 40–5 victory over Phrizbee in the opening round, and KO'd Tripulta Raptor in Round 2 before being eliminated via KO by Surgeon General in the third round.

KILLERHURTZ

A heavyweight built on wheels with battery-powered drive, KillerHurtz's main weapon is a compressed-gas-driven ax with interchangeable heads. That ax is driven by a pneumatic cylinder that produces almost a ton of force. Large four-liter buffer tanks maximize the speed of the ax. The chassis and the armor are comprised entirely of polycarbonate sheet. KillerHurtz travels at a top speed of sixteen miles per hour. Built by John Reid and Iona Foster, both of Oxford in the United Kingdom, KillerHurtz took almost three years to build.

They say the whole world loves a beheading. No, seriously, they do. Proof of that old axiom came at *BattleBots*' Season 1 heavyweight competition, when KillerHurtz earned a place on the all-time *BattleBots* highlight reel when it used its ax to knock the mannequin head off a foe named Knome II.

During Season 2, KillerHurtz was battered by Golddigger's swinging weapon and eliminated in the opening round.

At *BattleBots*' Season 3 tournament, KillerHutz KO'd Center Punch in the opening round, won a 30–15 decision over Omega-13 in Round 2, and was then defeated soundly by Hexadecimator in the third round.

At the *Robot Wars* 3 (UK) competition, KillerHurtz lost in the opening round of its heat to Cerberus.

KILLJOY

Built over a two-month period by Bob and Christina Knissel of Las Vegas, Nevada, Killjoy is a wedge-shaped bot. Its offensive strategies are twofold. It can use its wedge front to flip an opponent, with the help of its pivoting chassis, or its tail to ram.

KILL-O-AMP

Vince Lombardi once said, "Football is not a contact sport. Dancing is a contact sport. Football is a collision sport." Well, the coach might as well have been talking about bot fighting, and Kill-o-amp in particular.

A combatant in the heavyweight division, Kill-o-amp's main weapon is its own ability to dish out harder collisions than it has to absorb. Indeed, this wheeled and battery-powered bot, eight months under construction, has the ability to take out a competitor with one blow.

As main builder Curt Meyers of San Jose, California, points out, "A two hundred-pound bot traveling very fast can transfer a lot of energy in a collision."

In practice, however, things did not work out that well. In the opening round of *BattleBots*' Season 3 heavyweight tournament, Kill-o-amp was KO'd by the eventual champion, Son of Whyachi.

Meyers's teammate is Amy Sun.

KINETIC KILL VEHICLE

Alhambra, California's, Jason Gates is the creator of Kinetic Kill Vehicle, which was knocked out and eliminated by Timmy in the opening round of *BattleBots*' Season 3 middleweight tournament.

KING B3

At the *Robot Wars* 4 (UK) tourney, King B3 lost in the opening round of its heat to the unseeded Atomic.

KING BUXTON

King Buxton performed very well at the *Robot Wars* 3 (UK) competition. It defeated Eric in the opening round of its heat, bested Weld-Dor in the second round, and then lost in the heat finals to 101.

KNOME II

A participant in *BattleBots*' Season 1 heavyweight tournament, this bot started with a mannequin's head attached to it—you know, to give it a little personification. Knome II then earned a permanent place on the *BattleBots*' highlight reel when it lost its head to KillerHurtz's driving ax.

KRAKEN

The creation of James Moore and Team Zeus of Chico, California, Kraken advanced past Batray in the opening round of *BattleBots*' Season 3 middleweight tournament, then lost via KO to T-Minus in the second round.

KRAKEN (II)

This Kraken is made up of a stainless steel frame, and stainless steel, Kevlar, titanium, and neoprene composite armor. Neoprene, I have been led to believe, serves to absorb impacts.

Kraken operates with two 4-horsepower motors for drive, two twelve-ounce pneumatic motors for independent weapon control, nicad battery power, and controversial 4QD golf cart electronic speed controls. They are controversial because they were mounted externally. Kraken is painted to resemble a Gaelic sea serpent. It takes two to drive this monster.

It was built by Eric Koss, age thirty-three, and wife, Joyce Koss, age twenty-nine, from Virginia Beach, Virginia. They have been researching and developing in robotics for two years now. He is a reconnaissance instructor in the U.S. Navy. Other members of Kraken II's team include Chris Hinson, Phil Koss, and Jan Koss.

This Kraken was a competitor at *Robotica II.*

KRITICAL MASS

The techno-wizards behind Kritical Mass are Jeff Cesnik, age twenty-seven, of Stephens City, Virginia, and William "Billy" Boden, age twenty-eight, of Winchester, Virginia, who have been experimenting with robotics for six years.

Kritical Mass took three months and $7,000 to build. The builders, when they are not designing pugilistic gadgets, own and operate their own Internet consulting firm: Cesnik Technologies.

Kritical Mass weighs 201 pounds and is made of mild steel frame, mild steel armor, and 6061 aluminum bits and pieces. It has six-wheel drive and two-stage chain-driven speed reduction.

It is powered by two sealed lead-acid batteries and three electric motors. Its weapon boasts 3 horsepower! Its vertically mounted blade—The Spinning Blade o' Death, it is called—spins at 2,400 revolutions per minute!

If that does not do in its opponent, Kritical Mass also has a carbide-tipped ramming point.

In *Robotica*'s Season 1 tournament, Kritical Mass beat Grimlock in the preliminary rounds and then disposed of Hot Wheels in the Fight to the Finish. This landed Kritical Mass in the Final Six, where it was put up against Run Amok. This bot managed to win the Maze, but was defeated on the Speedway and Gauntlet. Though the final score was tied at 210, Run Amok advanced because it had won two of the three rounds.

KRITICAL MASS II

Like its predecessor, Kritical Mass II was built by Jeff Cesnik and Billy Boden. This bot, sometimes referred to by the abbreviation KMII, performed well at the *BattleBots'* Season 3 heavyweight tournament, winning a 38–7 victory over Spinne in the opening round and a 35–10 decision over Chewbot in Round 2. The bot was eliminated in the third round, losing a razor-close 23–22 decision to Golddigger.

KMII weighed 145 pounds and was powered by forty nickel-cadmium cells, five electric motors, and six-wheel drive. Its weapons were another "Spinning Blade o' Death" and a carbide rammer.

The bot was built of 4130 aircraft-steel frame, grade-5 titanium armor, and anodized 6061 aluminum bits and pieces. It took three months and $12,000 to build.

KRYPLER

Krypler, says builder Bob Knissel, age thirty-four, a management administration/account executive from Las Vegas, Nevada, "was designed specifically for *Robotica.* I built in two inches of ride height and used tank steering for quick manipulation. It also includes an eight-inch tri-chip carbide blade mounted verically and two six-inch spikes on opposing ends. I used diamond plate for the skin."

A previous effort of Knissel's appeared at the 2000 BattleBots competition in his hometown.

L

LAMACHINE

LaMachine was one of the first battling robots. It fought at the original San Francisco Robot Wars shows. The creators of LaMachine were Edward

The set for *Robot Wars: Extreme Warriors*. Let the rumble begin!
(Photo courtesy Brent Hunter/Team Boltz)

A bot's-eye view from the starting line of *Robotica*'s Gauntlet.
(Photo courtesy John Adipietro)

Mark Beiro and Tazbot.
(*Photo courtesy BattleBots, Inc.*)

Professor Noel Sharkey of Creative Robotics and friend.
(*Photo courtesy Christine Melia*)

A LEGO®-based robot navigates a tricky corner of the Acroname Obstacle Course. (*Photo courtesy of and © Jim Sincock/Acroname*)

A basic line-following robot made from a LEGO® Invention System navigates a track unique to Acroname.
(*Photo courtesy of and © Jim Sincock/Acroname*)

Manta has a carbide rammer as a front weapon and a spinning steel blade as a rear weapon. (*Photo courtesy Jeff Cesnik*)

Manta without its grade-5 titanium armor. (*Photo courtesy Jeff Cesnik*)

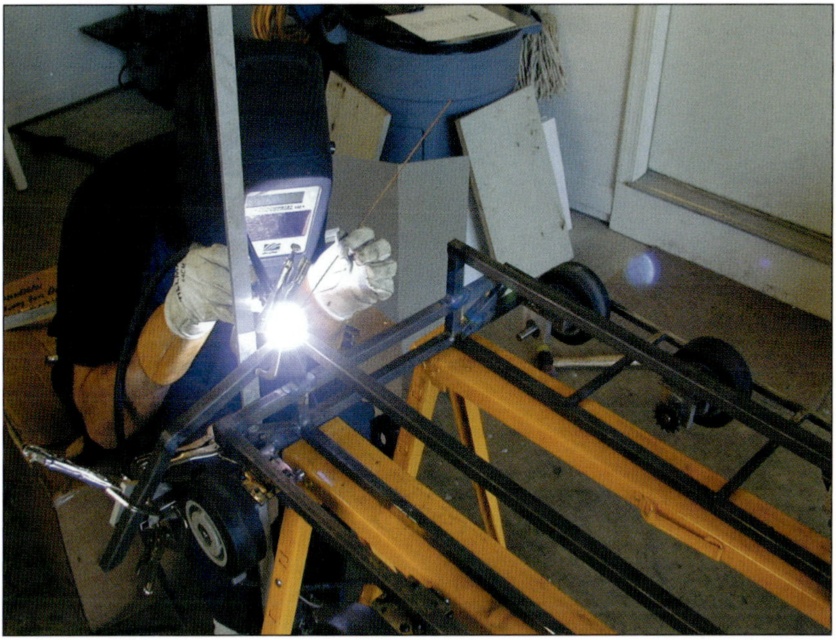

Manta's pieces were carefully soldered into place. (*Photo courtesy Jeff Cesnik*)

A spinning blade is securely attached to Manta's 4130 aircraft-steel frame. (*Photo courtesy Jeff Cesnik*)

The circulatory system of any bot is its wiring. (*Photo courtesy Jeff Cesnik*)

The frame must support the bot's weapons, power, and armor. (*Photo courtesy Jeff Cesnik*)

Manta derives its power from groupings of nickel-cadmium cells.
(*Photo courtesy Jeff Cesnik*)

Holes are carefully drilled so the armor can be attached to the chassis.
(*Photo courtesy Jeff Cesnik*)

Manta's armor is carefully fitted into place. (*Photo courtesy Jeff Cesnik*)

Team Suspect (*left to right*): David Gilson, Arthur Heckman, Jeff Cesnik, and Billy Boden at *Robotica* with UltraViolence ready for action. (*Photo courtesy Jeff Cesnik*)

JuggerBot, built by Nicole Morrow's husband Mike, performed well at *Robotica I*. (*Photo courtesy Jeff Cesnik*)

The bot Mad Cow with its wedge in front. (*Photo courtesy Jeff Cesnik*)

Mad Cow's steel frame resembles an erector set.
(*Photo courtesy Brent Hunter/Team Boltz*)

Brent Hunter and Jason Barger work on Mad Cow.
(*Photo courtesy Brent Hunter/Team Boltz*)

Team Boltz's newest bot is Psycho Chicken.
(Photo courtesy Brent Hunter/Team Boltz)

The 220-pound MiniRip was built in three weeks.
(Photo courtesy Chris Hannold/Litewav Robotics)

Mark Whitney's Nancy in the Trinity Fire Fighting maze.
(*Photo courtesy of and © Jim Sincock/Acroname*)

The Revolutionist is a fine example of a spinner. (*Photo courtesy Jeff Cesnik*)

The BrainStem-controlled Rover uses an IPaq with wireless LAN and camera for teleoperation. (*Photo courtesy of and © Jim Sincock/Acroname*)

Mark and Aaron Joerger with Run Away, their 218-pound wedge-shaped bot. (*Photo courtesy Mark Joerger*)

Sobek is a veteran of *Robot Wars: Extreme Warriors*.
(*Photo courtesy Jeff Cesnik*)

Son of a Monkey Wrench on the starting line.
(*Photo courtesy John Adipietro*)

Team Rous tests its bot in a parking lot.
(*Photos courtesy John Adipietro*)

Son of a Monkey Wrench on its way to the *Robotica* show.
(*Photo courtesy John Adipietro*)

Team Rous (*left to right*): John Adipietro, James Uhlinger, and V. J. Petker before a match. (*Photo courtesy John Adipietro*)

"Trey" Roski III and his cousin, Greg Munson of Oakland, who went on to start the *BattleBots* TV show.

LAMBO

Built by Scott Gary and Andrew Gardner of Santa Barbara, California, Lambo may not be your worst nightmare, but it is a lammer . . . er, rammer, to be reckoned with. This lightweight bot runs on batteries and wheels.

LAND MINE

Team TNT2 and main builder Mark Talcott of Torrance, California, built Land Mine, which lost a 26–19 decision to Vermicious Kenid in its first bout at *BattleBots*' Season 3 lightweight tourney.

LICENSE TO KILL

License to Kill was constructed by Valek Sykes and S.P.E.C.T.R.E. Robotics of Valencia, California. The bot was knocked out by Bacchus in the opening round of *BattleBots*' Season 3 heavyweight tournament.

LI'L CHIPPER

Li'l Chipper was built by Scot Martin and Team Playa of Hoboken, New Jersey. It KO'd Avatar in the opening round of *BattleBots*' Season 3 middleweight tournament before losing by KO to F5 in Round 2.

THE LITTLE BLUE ENGINE THAT COULD

With one of the longest names in fighting robot history, The Little Blue Engine That Could was both constructed and named by Josh O'Briant of San Diego, California. This bot lost in the opening round of *BattleBots*' Season 3 superheavyweight tournament—but not by much. It fell on the short end of a 23–22 decision to Techno Destructo.

LITTLE DRUMMER BOY

This bot excelled at *BattleBots*' Season 3 middleweight tournament. Built by Steve Buescher and Team Dangerous Drums of Pembroke Pines, Florida. This bot emerged victorious in seven consecutive bouts.

Little Drummer Boy defeated Short Circuit by a score of 41–4, KO'd Flip You Off, outpointed Ice Cube 29–16, then KO'd Bouncing Betty, Blade Runner, and Eraser in that order. It defeated Sabotage in the semis and lost to Hazard in the finals.

The bot has a simple design and uses a powerful vertical spinning weapon to take out its opponents. The bot uses two 24-volt DeWalt Hammerdrill motors to spin a twenty-five-pound steel drum.

The weapon can achieve an impact force of eighty-five miles per hour. Two ½-inch square bars are symmetrically welded onto the drum. By building a wider chassis and a narrower spinning weapon, Buescher does not have the same problem with gyroscopic procession as other spinning bots.

LITTLE EGYPT

Little Egypt does not have a very intimidating name. Its combat record is not going to impress anyone either. In the opening round of *BattleBots*' Season 3 tournament, this superheavyweight—created by Jim Kennon of Benicia, California—lost a 39–6 decision to Eradicator in its opening bout.

LITTLE SISTER

Because of her name, folks tend to underestimate Little Sister. But there is nothing little, or feminine, about this bot, which was built by Ian Watts and Team Big Brother of Brighton, Sussex, United Kingdom. In the opening round of *BattleBots*' Season 3 heavyweight tournament, Little Sister won a near shutout (43–3) over Botknocker 19. In Round 2, Little Sister put a big hurt on Gammatron and won by KO. BioHazard then KO'd Little Sister in the third round.

LIQUID LUNCH

More dangerous than three martinis before noon, Liquid Lunch is a wheeled bot that runs on both battery power and hydraulic fluid. Its weapon, a crushing device, is totally hydraulic.

The wedge-shaped bot has two-wheel electric drive, a 13-horsepower gasoline engine, and a large two-stage log-splitter hydraulic pump. This bot was built by the Nelson family of Sparks, Nevada, with Steven being the main builder, while Martha and Lowell rounded out the team.

LOBOTOMIZER

Michael Glenn of Portola, California, built Lobotomizer, which—instead of aiming an ice pick directly for the inside of your eyelid—was knocked out by Wedge of Doom in its first match at *BattleBots*' Season 3 lightweight tourney.

LOBOTOMY

Israel Mathewson and Team Gray Matter of Medford, Oregon, constructed Lobotomy, which outpointed Maddgoth by a score of 40–5 in the opening round of *BattleBots*' Season 3 middleweight tournament, then defeated Twister in Round 2 by a score of 32–13, but lost a 25–20 decision to Twin Paradox in the third round. Kind of makes your frontal lobe ache just to think about it.

LUG NUTS

Built in two weeks by Shawn Ouderkirk of Colorado Springs, Colorado—with the help of teammates Tony Aug and Jeremy Graham—Lug Nuts is a wheeled bot powered by compressed gas.

M

MACE II

In the *Robot Wars* 3 (UK) tournament, Mace II performed very well, winning its heat with consecutive victories over Stinger, Weeliwako, and Suicidal Tendencies. In the semifinal heat, Mace II defeated Big Brother in the opening round, but failed in its attempt to make it into the tournament's final four when it was eliminated in the semifinal heat's second round by the eventual overall winner, Chaos II.

MACHIAVELLI

Hey there, all you control freaks, here's the bot for you. Machiavelli, named after a puppetmaster who preferred his puppets human, was birthed by Christo Logan and Team Fear of Berkeley, California. It then lost control and was killed by Orlando Commando in the opening round of *BattleBots*' Season 3 middleweight tournament.

MAD COW (I)

Mad Cow was KO'd of the opening round of *BattleBots'* Season 3 lightweight tourney by Vermicious Kenid. It was built by Justin Morgenthau of Manchester, Connecticut.

MAD COW (II)

This Mad Cow was built by Brent Hunter, Jason Barger, and Team Boltz of Los Angeles, California. It is powered by a 24-volt battery and has a top speed of fourteen miles per hour. Mad Cow's weapons are a spiked ram wedge and a rear saw.

It is armored with steel, plexiglass, and wood. Mad Cow, which was adapted from this team's previous bot, Boltz, competed in *Robot Wars* in London in June and July 2001. Team Boltz dresses as farmers during their bots' matches.

About Team Boltz, Brent Hunter says, "We're both a couple of hobbyists that don't really know much about mechanics, but have seemed to find a knack for putting robots together from mostly hardware-store parts. It all just kind of makes sense.

"I never thought eighth-grade metal shop would come in handy. I won a couple of LEGO Mindstorms' expert division competitions with my own LEGO creations—so it only seemed natural to move on to metal. I think we learned the bulk of mechanical engineering from LEGO."

Word is that this bot is again being revamped, this time as Psycho Chicken. That way, Team Boltz gets to reuse the farmer outfits.

See also BOLTZ.

MADDGOTH

As if there's any other kind. Maddgoth was another short-lived bot. Built by Bill Bottenberg and Team Crash of Everett, Washington, Maddgoth was thumped 40–5 by Lobotomy in the opening round of *BattleBots'* Season 3 middleweight tournament.

MAGGOT

Gary Warren of Blackfoot, Idaho, built Maggot, which was quickly eliminated from *BattleBots'* Season 3 middleweight tournament. Maggot lost a 28–17 decision to F5 in the opening round.

MAJOR TOM

At the *Robot Wars* 4 (UK) tourney, Major Tom, much in need of ground control, was defeated in the opening round of its heat by 101.

MALICE

Lionel Vogt and Team Violence of Dunwoody, Georgia, built Malice, which won via KO in the opening round of *BattleBots'* Season 3 lightweight tourney before losing, also by KO, to Carnage Raptor in Round 2.

MALVOLIO

As, of course, you already know, Malvolio is named after a character in Shakespeare's play *Twelfth Night.* This Malvolio is a wedge, twenty-four inches long by eighteen inches wide. It stands ten inches tall and tips the scales at a speedy 90 pounds.

Malvolio is powered by seventy-two nicad cells. Its two drive wheels are driven by EV Warrior Electric Bike motors and uses tank-style steering.

A titanium bar provides the weapon. Tied to each end of the bar are one-pound weights. Then the bar spins around at 1,000 revolutions per minute. Ouch. Another EV Warrior motor drives the weapon, which is capable of delivering two kilojoules of energy on impact.

The bot's frame, which is painted yellow, is made from ¾-inch square tubing and is armored with aluminum, Lexan, and titanium. The builder is Ramon Ebert, a thirty-four-year-old software engineer for Sony Electronics. Malvolio and his builder made their debut at the 2001 *Robotica.*

At *BattleBots'* Season 3 middleweight tournament, Malvolio was outperformed by Iron Soldier, losing a 41–4 decision.

MANEATER

Here's a bot from my neck of the woods: Brooklyn, New York. Maneater was built by Amy Sperber and her "Brooklyn Powers." Maneater lost a tough fight 25–20 to Eye of the Storm in the opening round of *BattleBots'* Season 3 middleweight tournament.

MANIC AGGRESSIVE

Eric Sporer and Team Final Jeopardy of Brooklyn, New York, built Manic Aggressive, which outpointed Tantrum 27–18, KO'd Mantis, then lost by a 30–15 decision to Serial Box Killer in *BattleBots*' Season 3 lightweight tourney.

MANIC MUTANT

At the *Robot Wars* 3 (UK) competition, Manic Mutant was defeated by Gravedigger in the opening round of its heat.

MANTA

Manta weighs 185 pounds and measures forty-four inches by thirty-seven inches by thirteen inches. It is powered by eighty nickel-cadmium cells, five electric motors, and six-wheel drive. Its weapons are a spinning steel blade, carbide rammer, front wedges, and, according to the builders, a "killer paint job."

Manta is quick and powerful but lacks the ability to self-right. It was built in two months by Jeff Cesnik, Billy Boden, and Chris Tyson from Virginia, using a 4130 aircraft-steel frame, grade-5 titanium armor, and anodized 6061 aluminum bits and pieces.

In the "Civil War" episode of *Robot Wars: Extreme Warriors,* Manta represented the South and ended up winning the whole ball of wax. It KO'd Tiger Cat, Spike, and, in the finals, The Six Million Dollar Mouse, to win the war for the "land ub cotton."

MANTARAY

Built by Ted Shimoda of Pasadena, California, Mantaray lost in the opening round of *BattleBots*' Season 3 superheavyweight tournament. Mantaray was knocked out by Phere.

MANTIS

Definitely not the praying kind, Mantis was built by Jonathan Bayles and Team Spike of Somerville, New Jersey. In *BattleBots*' Season 3 lightweight tourney, Mantis outpointed Speed Bump decisively (36–9), KO'd Frogemoth, then lost via KO to Manic Aggressive.

MASSACRE

Built by Daniel Rupert and Team Malicious of San Diego, California, Massacre was KO'd in the opening round of *BattleBots*' Season 3 heavyweight tournament to BattleRat.

THE MASTER

The Master runs on battery power and on wheels. Its primary weapon is a gas-powered cutoff saw and actuated lifting arm. This middleweight bot, a veteran of *BattleBots,* was built in a month by Mark Setrakian, Paul Rivera, and Peter Abrahamson of Los Angeles, California.

MATILDA

Matilda is one of the house bots for *Robot Wars: Extreme Warriors*—one of the bots that patrol the arena and rub salt in the wounds of all the bots who are getting the worst of it. Matilda weighs 256 pounds and measures fifty-six inches by twenty-seven inches by twenty-seven inches.

It is powered by a prototype electric motor and is well armored. It looks like a cross between a triceratops and an armadillo and has two effective weapons: tusks, which can impale an opponent, and a sixty-pound flywheel.

MATTS BAMMER

Jeff Vasquez and Fast Electric Robots of Oak Park, California, built Matts Bammer, which advanced to the fourth round of *BattleBots*' Season 3 lightweight tourney. It shellacked Robo Master 41–4, outpointed Double Secret Probation 34–11, and coldcocked Rhode Hog before being eliminated via KO by Backlash.

MAULER 5150

Mauler 5150 features rotary impact weapons on a lid that spins at 550 revolutions per minute. The spinning weapons include two spiked steel flails and two steel attack chisels. The wheeled and battery-powered bot also features a 3.5-horsepower spinner motor.

The bot has evolved over several different versions since it made its combat debut in 1995. The current version is more than four times more powerful than the original.

The main builder is Charles Tilford of Portola Valley, California. The rest of his team is comprised of Morgan Allen-Tilford, Kim Knollenberg, and Alan Vermette.

At *BattleBots*' Season 2 heavyweight tournament, Mauler 5150 was flipped onto its back and eliminated in the opening round by Bigger Brother. In *BattleBots*' Season 3 tournament, Mauler 5150 knocked out Dreadbot in the opening round before losing 25–20 to Incoming in Round 2.

MAX DAMAGE

Max Damage was a loser in the opening round of its heat at the *Robot Wars* 3 (UK) tournament, eliminated by Agent Orange.

MAXIMUM OVERDRIVE

Constructed by Jeff Mills and Team Overdrive of Edwardsville, Illinois, Maximum Overdrive was KO'd in the opening round of *BattleBots*' Season 3 heavyweight tournament, punctured by the fearsome fangs of Incisor.

MAXIMUS

Superheavyweight Maximus was designed and built by Pat Lidstrom of Los Angeles, California. The bot lost a 31–14 decision to Hammertime in the opening round of *BattleBots*' Season 3 tournament.

MAZAKARI

At the *Robot Wars* 4 (UK) tourney, Mazakari lost in the opening round of its heat to Iron Awe.

MECHACIDAL MANIAC

With four-wheel drive and four-wheel steering, Mechacidal Maniac has a power plant capable of generating 12 horsepower and can get up to seventeen miles per hour. More impressively, it can accelerate from zero to seventeen in two seconds.

Mechacidal Maniac was built by Brian Nave, age thirty-eight, of Ormond Beach, Florida, who has been experimenting in robotics for thirteen years. That longevity of interest comes because for Brian, robotics is not a hobby. It is his profession. He designs and builds industrial robots for a living, so building a combat bot comes second nature.

Mechacidal Maniac made its TV debut at *Robotica II* in 2001. Also on the team is Michael Nave.

MECHADON

Winner of the Medal of Honor at the *BattleBots*' Season 1 superheavyweight tournament. Although not the greatest fighter, Mechadon walks in a spiderlike fashion—with style and grace on huge razor-sharp blades.

Mechadon did win one bout when Prompt Critical broke down, allowing the big spider to move in for the kill.

MECHANICAL WILDCAT

Mechanical Wildcat was constructed by Andrew McCutchen and the ASME Wildcats of Tucson, Arizona. More like a Mechanical Kitten, it was KO'd in the opening round of *BattleBots*' Season 3 middleweight tournament by Pokey.

MECHA TENTOUMUSHI

The creation of Lisa and Becky Winter and Robot Action League of Walnut Creek, California, Mecha Tentoumushi outpointed The Archduke 31–14 in the opening round of *BattleBots*' Season 3 lightweight tourney, then lost a 23–22 squeaker to Wedge of Doom in the second round.

The bot, which took six months to build, is battery powered and runs on wheels. Its weapons are its capturing dome and grappling hook.

MECHAVORE

A much-feared metal-eater, Mechavore was created by Robert Marzinske and Team Shrapnel of Tempe, Arizona. Mechavore was extremely impressive in *BattleBots*' Season 3 heavyweight tournament.

After winning a close 23–22 decision over El Diablo Grande in the opening round, it advanced past Technofool and scored an upset victory over Vlad the Impaler in the next two rounds. Mechavore scored a KO win over Surgeon General in the fourth round before finally losing to the eventual tournament champion, Son of Whyachi.

MEDUSA 2000

At the *Robot Wars* 4 (UK) tourney, Medusa 2000 defeated Atilla the Drum in the opening round of its heat, then lost in the second round to the tournament's defending champion, Chaos II.

MELTDOWN

Built by Tom Wickey and Bots of Wrath of Galveston, Texas, Meltdown was liquidated by Exodus 2001 in the first round of *BattleBots*' Season 3 middleweight tournament, losing a 32–13 decision.

METALHEAD

Back in my day a Metalhead was a guy who listened to way too much Black Sabbath, but today a Metalhead actually has a metal head. Designed and constructed by John Bamberg of Santa Rosa, California, Metalhead lost 32–13 to Techno Destructo in the opening round of *BattleBots*' Season 3 superheavyweight tournament.

METALLIC MAYHEM

Less the name of a robot than a perfect description of the entire sport, Metallic Mayhem was built by Daniel Rupert and Team Malicious of San Diego, California. The bot lost 30–15 to Sunshine Lollibot in the opener of *BattleBots*' Season 3 middleweight tournament.

METALMORPHIS

Metalmorphis is a thirty-five-inch square that is three inches high. It has an aluminum frame, in which is mounted three modular two-wheel drive/controller/battery/radio chassis—giving it a six-wheel-drive unit.

This bot earned its name when it competed at *Robotica II* in 2001, as it came out in three different configurations for the three challenges. Its wheels, bumpers, and weapons can all be changed in a flash to best meet the circumstances.

It was built by Les LaZar, age fifty-one, Dan Fetterly, age fifty-one, and Roberta Dwoskin, of Los Angeles, California. Les says that his favorite movie robots are Robby the Robot from *Forbidden Planet* and Gort from *The Day the Earth Stood Still.*

Dan is the president and CEO while Les is the chief design engineer for United Service Associates, Inc.

M.I.A. (MISSING IN ACTION)

Built by Steven West and Mark Lewis of Pleasanton, California, M.I.A. is battery powered and built on wheels. The middleweight, wedge-shaped bot took three months to design and two months to build. Known for its speed, M.I.A. likes to hit and run. Try and retaliate and poof! It's Missing in Action.

THE MIDNIGHT PROJECT

Ben Wright and Team Midnight of San Bruno, California, burned the midnight oil to build this bot. Perhaps they should have gotten a good night's rest. Their bot was KO'd by Wacker in the opening round of *Battle-Bots'* Season 3 lightweight tourney.

MILLY ANN BUG

Milly was defeated by a fellow insect, Bumble Bot, in the opening round of the *Robot Wars* 3 (UK) tournament. At the *Robot Wars* 4 (UK) tourney, it defeated Reptirron in the opening round of its heat, but was defeated in the second round by Razer.

MING

Named after the dynasty, Ming was defeated in the opening round of its heat at the *Robot Wars* 3 (UK) competition.

MING 2

At the *Robot Wars* 4 (UK) tourney, Ming 2 defeated Diotoir in the opening round of its heat, then was defeated in the second round by The Morgue.

MINI INFERNO

Don't let the size fool ya. Mini Inferno may not be the biggest bot in the Box, but it can do some damage. Most of Mini Inferno's 37-pound weight is in its drivetrain. That's because the body and frame are ultra-lightweight, made of a carbon-fiber composite and aluminum. The batteries, motors, and electronics make up most of the weight.

Mini Inferno's surprisingly strong drivetrain is made up of four power-drill motors and gearboxes—which were originally on 18-volt power drills—two 12-volt, 7ah batteries wired in series, and a pair of electronic speed controllers.

The bot has hinges on all sides. To these hinges can be bolted carbon-fiber flaps, which can be used either defensively, as a skirt, or offensively, as a scoop. To further menace his opponents, Mini Inferno has two removable hardened-drill-rod ramming spikes.

Mini Inferno is the creation of Jason Dante Bardis, a thirty-year-old graduate student pursuing his Ph.D. at the University of California, Santa Barbara, who has been tinkering with robotics for almost seven years.

Bardis's creations have competed at Robot Wars '96; Robot Wars '97; the E3 LEGO Mindstorms RoboGladiators, where he won third place; BotBash '99, another third place; BattleBots Long Beach '99; BattleBots Las Vegas '99; BotBash 2000, where he took second place; BattleBots San Francisco 2000, and BattleBots Las Vegas 2000.

MINION

Minion fights in *BattleBots'* superheavyweight division, weighing a hefty 325 pounds. Although driven by battery power, its primary weapon is powered by liquid fuel. That weapon consists of a row of all-purpose fourteen-inch-diameter cutting wheels driven by a 5-horsepower gas engine. Minion's other special features include an unbreakable drivetrain and a Lexan "tongue of wedge-death."

Christian Carlberg of North Hollywood, California, built Minion. His team is called the Cool Robots, and the other members are Brian Roe, Luke Khanlian, Jessica Lehrbaum, and Jason Lehrbaum. (CoolRobots.com is one of the best. Check it out.)

At *BattleBots'* Season 1 superheavyweight tournament, Minion won the Giant Nut, defeating Gray Matter in the opening round (in a match in which Minion lost its saw during the bout but emerged victorious anyway), Grendel in the quarterfinals, Rammstein in the semis, and DooAll in the finals.

Season 2 was a bad one for Minion. That season, Minion was eliminated—flipped onto its back and rendered unable to continue—in the opening round by Atomic Wedgie.

In *BattleBots'* Season 3 tournament, Minion won four matches before being eliminated from the competition in the finals. In the opening round, Minion KO'd Tortoise. It followed up that victory with a 23–22 points win over Ogre, a 24–21 victory over Electric Lunch, and a 27–18 win over Toro. In the fifth round, however, Minion met its match and lost a 26–19 battle to Vladiator in the finals.

MINIRIP

MiniRip is the creation of Chris Hannold and Litewav Robotics of Salisbury, North Carolina. A heavyweight, it weighs 220 pounds. It receives its power from four 24-volt electric motors. It features a 10.5:1 gearing ratio and rolls around on 10-inch knob tires.

Its primary weapon is a hammer disk that spins at 2,500 revolutions per minute, and it can also hurt its opponent with its 30-inch steel bulldozer blade. For protection, MiniRip wears ¼-inch 6061 aluminum and ½-inch Lexan.

It is radio controlled via Hitec seven-channel PCM remote and four IFI Victor Speed Controllers.

MiniRip was built in three weeks to compete in *Robotica*'s second season. It was a last-minute replacement after many entrants dropped out following the September 11 attacks on the World Trade Center and Pentagon.

To prove that Chris believes in recycling, here's his description of how MiniRip was built: "I decided that we didn't have enough time to fab everything from scratch. There were also some improvements that I wanted to make to other things we'd done in the past. I still had the aluminum body of RipOff that I wasn't planning to use in the rebuild. So I cleaned it out and prepared it for MiniRip. I didn't think we'd milked all the power out of the four EV Warrior motors and gearboxes that we'd used in HandsOff! and the Six Million Dollar Mouse. Guess what, they fit perfectly inside the old RipOff body and left plenty of room for batteries, control electronics and whatever else."

MISANTHROBOT X2

Erik Reikes and Jon O'Brien of Solana Beach, California, built this bot, which exited not long after it entered *BattleBots*' Season 3 lightweight tourney, KO'd in the opening round by No Tolerance III. Battery powered and built on wheels, its primary weapon is its wedge shape and its "reversing high-speed blade of death." Misanthrobot X2 took four months to build.

MISS ILE

Most maidens aren't named after, not to mention shaped like, phallic symbols—but Miss Ile is! The unmarried bot lost her debut to Anorakaphobia in the opening round of the *Robot Wars* 3 (UK) tournament.

MISSING LINK

At *BattleBots'* Season 1 lightweight tournament, Missing Link was eliminated in the opening round by future champion Ziggo.

MR. BONESTRIPPER

Bots often have names that sound like the titles of 1981 slasher movies. Such is the case with Mr. Bonestripper, a wheeled bot with battery-powered motor and a liquid-fuel-powered-main weapon. That weapon consists of counter-rotating wheel blades.

Builder Ray Scully of San Anselmo, California, who calls this bot "scary ideas with substandard construction," put Mr. Bonestripper together in three weeks.

Scully's teammate is Joshus Ogle. Scully loves to sing this bot's praises: "There are at least four welds held together with duct tape," he says. "The electrical systems are melting as we speak."

MR. PUNCH

At the *Robot Wars* 3 (UK) tournament, Mr. Punch was eliminated in the opening round of its heat by the cruel justice of (insert sound of three gavel raps) Judge Shred.

MR. ROBOTO

Mr. Roboto, a wedge-shaped bot that is battery powered and operates on wheels, is a pure rammer and claims as one of its strengths its agility. This middleweight is a veteran of *BattleBots.* It was constructed by Kyle Hickerson, Phillip Perez, and William Tsun of Las Cruces, New Mexico.

MR. SNOOGGLEPUSS

A lightweight bot out of Delafield, Wisconsin, Mr. Snoogglepuss was constructed by Jonathan Stoffer and his team of "Scroungers." It was outpointed 28–17 by Bambino in the opening round of *BattleBots'* Season 3 lightweight tourney.

MR. SUCKO

Do not confuse this bot with Mankind's former tag-team partner. Mick Foley's friend was named Mr. Socko. They are two entirely different entities. Mr. Sucko was built by Nick Williams and Team Destro of Livermore, California. It was KO'd by Timmy in the opening round of *BattleBots*' Season 3 middleweight tournament.

MJOLLNIR

Fighting in the heavyweight division is the wheeled and battery-powered Mjollnir, which does most of its stinging like a bee—that is, hind end first. Mjollnir's primary weapon is a spiked-hammer tail. It has heavy steel construction and took four months to build.

Don Lariviere of Redondo Beach, California, built this bot with help from Annette Hetherington and Chris Myers. In *BattleBots*' Season 3 heavyweight tournament, Mjollnir was KO'd in the opening round by Surgeon General.

MOBSTER

Larry "Walnuts" Magnoli and Team Mag of Antioch, California, built the hitbot known as Mobster, which was rubbed out in the opening round of *BattleBots*' Season 3 middleweight tournament by Thwak (rhymes with whack).

M.O.E.

Any bot that is named after a Stooge is OK by me. Can a bot named Shemp be far behind? M.O.E. (which actually stands for Marvel of Engineering) is a wheeled and battery-powered bot. Its primary weapon is a high-impact spiked arm, which is CO_2 powered.

The Muzzy family from Cedar Falls, Iowa, is responsible for M.O.E., which took six weeks to build. The leader of the team is Neal, with Peter, Norman, and Melissa Muzzy rounding out the team. At *BattleBots*' Season 2 heavyweight competition, M.O.E. advanced as far as the third round, when it was eliminated by the eventual winner, BioHazard. In *BattleBots*' Season 3 tournament, M.O.E. lost to OverKill by a score of 28–17.

MORDICUS

Barry Mullins, who attends the U.S. Air Force Academy, built Mordicus, which lost 38–7 in the opening round of *BattleBots*' Season 3 middleweight tournament to the kung-fu bot, Crouching Puppy, Hidden Kitten.

THE MORGUE

At the *Robot Wars* 4 (UK) tourney, The Morgue defeated Scar in the opening round of its heat, defeated Ming 2 in the second round, then lost in the heat finals to Fire Storm 2.

MORTIS

The battery-powered Mortis moves on tracks and has two main weapons: a sledgehammer-style piercing blade that can smack a foe at a rate of three times per second, and a flipping arm, which gets under its opponents and flips them over onto their backs. That arm also can be used to put Mortis right-side up if he gets flipped over.

The bot features Kevlar/carbon armor and has a three-phase DC brushless servo that drives at 1200 watts each. It also has traction control with optical encoder feedback from the drive motors. Mortis took nine months to build.

The main builder is Rob Knight. His teammates are Arthur Chilcott and Ben Gordon. Mortis defeated Ming in the opening round of its heat at the *Robot Wars* 3 (UK) competition, and was defeated in the second round by Gravedigger. At the *Robot Wars* 4 (UK) tourney, Mortis defeated Crusader 2 in the opening round, then lost in the heat finals to Steg 2.

MOUSER MECHA-CATBOT

With a mystical name that sounds as if it were translated poorly from the Japanese, Mouser Mecha-Catbot was built over a six-month period by Fon Davis and April Mousley of San Francisco, California. It takes a lot of confidence to paint your bot pink. In *BattleBots*' Season 1 lightweight tourney, Catbot outpointed Shaft before being eliminated. Battery powered and built on wheels, Mouser Mecha-Catbot's primary weapon is a pneumatic flipper and tail.

MOUSER SUPER MECHA-CATBOT

Fon Davis's follow-up to Mouser Mecha-Catbot. With feline fury, this Super Catbot KO'd Blood Moon in the opening round of *BattleBots*' Season 3 lightweight tourney, KO'd Trilobot in the second round, and then was KO'd itself in Round 3 by Gamma Raptor.

MUSASHI

Musashi leveled Vortex in the opening round of *BattleBots*' Season 3 middleweight tournament, then was eliminated in Round 2 via a 34–11 decision by Short Order Cheff. It is the creation of Evan White and Team Kaatana of Orange, California.

MY SON

The father of My Son is Keith Gresham of Vista, California, who built the bot with the help of Team Paragon Computers. At *BattleBots*' Season 3 heavyweight tournament, My Son KO'd Piecemaker in the opening round before losing, also by KO, to Crab Meat in Round 2.

N

NAPALM

Oh man, you haven't lived until you've smelled this bot in the morning! Smells like victory! Napalm performed well at the *Robot Wars* 3 (UK) competition, defeating Kater Killer and Robo Pig before losing in the finals of its heat to Steg-a-Saw-Us.

NARWAL

Scott Kilbourne and The Wrecking Crew of Lewisville, North Carolina, created Narwal, a bot that lost a 35–10 decision to Deb Bot in the opening round of *BattleBots*' Season 3 middleweight tournament.

NASTY OVERBITE

Don't call this bot Bucky! Built by Tim Berghoefer and Team Jawbreaker from New Berlin, Wisconsin, Nasty Overbite KO'd Total Chaos in the opening round of *BattleBots*' Season 3 heavyweight tournament, then lost a 24–21 decision to B.O.B. in Round 2.

NEB

Karen Sharp and "Nebecannezer" of Mesilla Park, New Mexico, constructed Neb, which was KO'd by B.E.H.emoth in the opening round of *BattleBots*' Season 3 middleweight tournament.

NECRONOMIBOT

Erik Reikes and Team Misanthrobotics of San Diego, California, built Necronomibot, which was KO'd in the opening round of *BattleBots'* Season 3 middleweight tournament by Travesty.

NEGATIVE REINFORCEMENT

Constructed by Glen Sears and Hunter Robotics of Woburn, Massachusetts, Negative Reinforcement lost in the opening round of *BattleBots'* Season 3 heavyweight tournament to Slam Job by a score of 33–12.

NEMISIS

Nemisis was rendered motionless by Sunshine Lollibot in the opening round of *BattleBots'* Season 3 middleweight tournament. It was the creation of David Hall and the Silicon Valley Destruction Company of San Jose, California.

NEPTUNE

Built by Simon Arthur of Doraville, Georgia, superheavyweight Neptune lost by KO to JuggerBot in the opening round of *BattleBots'* Season 3 tournament.

NERD KILLER

Built by Casey Bennett and Team SMC of Santa Barbara, California, Nerd Killer may be able to efficiently dispose of anyone with a pocket protector and glasses taped together at the middle, but it struggled with Towering Inferno, losing by a 23–22 score in the opening round of *BattleBots'* Season 3 heavyweight tournament.

NICE BOT

Jason Crum and Team Nice Guys of Fairborn, Ohio, built Nicebot. I wanted to think that the name was ironic but Nice Bot was thrashed nicely and KO'd by Aggravated Assault in the opening round of *BattleBots'* Season 3 lightweight tourney.

NIGHTMARE

Nightmare's primary weapon is a vertically mounted spinning disc in front. The disc is not a saw blade but rather what builder Jim Smentowski calls "a rotary bludgeoning device," which, to give it bite, has had two large stainless steel teeth, on opposite sides, bolted securely to it. The disc could not be more damaging even if it were a saw blade, as it can tear an opponent up when it begins to spin at its top speed of 300 miles per hour.

The bot has small feet in front with casters that do not actually ride on the ground. They do have a purpose, however, as they keep the spinning disc from touching the floor when it misses its target. They also add stability to Nightmare.

Nightmare may not be the fastest bot on the block, but it is one of the strongest. The bot can easily tow a car, and usually emerges victorious when it gets into a shoving contest with a competing bot.

The 11.5-inch-diameter tires have a custom-made hub attachment to the shaft of the motor with a taper bushing. Nightmare's design is actually very simple: A 6061 aluminum frame holds the bot together and a small sheet of titanium provides the electronics with the only armor it needs.

Smentowski says that one of Nightmare's strong points is its welds. To keep it from coming apart he decided not to do the welding himself, but rather at a local welding shop. Jim says he was "making sure that there was *no* weak points that would allow breakage or flexing. I admit I used some serious overkill on the strength of this frame, but I wanted to take no chances."

Because of the strength of the frame, Jim drilled holes at the robot's rear to reduce weight. One PC545 Hawker Odyssey Drycell battery powers Nightmare's weapon and a second one operates the drive motors.

In *BattleBots*' Season 2 heavyweight competition, Nightmare was defeated in the second round by the eventual winner, BioHazard. At *BattleBots*' Season 3 tournament, Nightmare KO'd Slam Job in the opening round, then lost to the eventual winner of the tournament, Son of Whyachi, in Round 2.

NO APOLOGIES

The brainchild of John Torkelson of Kent, Washington, No Apologies KO'd Isosceles 1 in the opening round of *BattleBots*' Season 3 tournament, but then was itself KO'd by The Judge in the second round.

NOLL

According to Noll's main builder, twenty-four-year-old David "Django" Dexter, "We took the hands-on approach in building Noll. Every piece of metal was hand-cut and fabricated to do a specific job. We weren't worried about the outer appearance of our bot."

This bot never received a paint job and there is no covering for decorative purposes. In other words, everything on Noll has a purpose, and that purpose never, ever has anything to do with femmy aesthetics.

The hammer weapon has a seven-inch jackhammer spike that travels 200 degrees in 0.15 seconds. The lifting forks on the front can lift 950 pounds. Titanium sheeting from an Apache helicopter was used to protect the bot's vital organs.

Noll has the unique feature of ride-height adjustability to enable clearance over large obstacles.

Dexter's bot-building teammates include Mike Stefani, age twenty-four, Jed Saunders, age twenty-two, Kimberly Morcaldi, age twenty-four, Nicole Lowe, age twenty-four, Abraham Fechter, age twenty-one, and John Geurra, age thirty; all are from San Luis Obispo, California.

They are also all students at California Polytechnic State University. Noll is this team's first attempt at bot building. Noll made his battling debut on the first series of the *Robotica* TV show.

NORTHERN FURY

Maneuverability is the name of the game with Northern Fury, a name so strong that it ought to belong to a thoroughbred racehorse. This bot, which comes from north of the border, up Canada way, is a two-wheeler with a frame built of high-strength steel, protected on all sides by aircraft-grade aluminum panels.

Locomotion is provided by two high-power wheelchair motors, one for each wheel. Its front wedge sits on the ground while its hind end is jacked up, so it has the best of both worlds, low clearance for combat and high-clearance for getting past obstacles. The high-clearance end has twin chain saws that are designed to lift the opponent off the ground rather than slice 'n' dice.

Northern Fury was built by Neale McDavitt-Van Fleet, age twenty, Charles Halliburton, age twenty, Mac Larwill, age twenty, Scott Moreland, age seventeen, and Ken Hutchison, age fifteen, of Ottawa, Ontario,

Canada. They were champions of the 2001 Canada First Robotic Games, a nationwide competition for high school students.

Northern Fury made its U.S. debut at *Robotica II.*

NO TOLERANCE II

Stefan Nock and his No Tolerance Combat Robotics organization of Baltimore, Maryland, are responsible for No Tolerance II, which won a 32–13 decision over Kill-9 in the opening round of *BattleBots'* Season 3 middleweight tournament before losing a 25–20 decision to Ravager in Round 2.

NO TOLERANCE III

The lightweight entry from Stefan Nock, this bot lost in the opening round of the *BattleBots'* Season 2 lightweight competition to Toe Crusher. It KO'd Misanthrobot X2 in the opening round of *BattleBots'* Season 3 contest, then came up on the short end of a 26–19 decision to Trilobot in Round 2.

It is a stomp bot that walks on six legs. It runs on electricity and its primary weapon is a battery-powered saw. Nock had help building and operating No Tolerance III from Trevor Bludis of Baltimore, Maryland.

NSYNCERATOR

Like the "boy" group it's named after, Nsyncerator has no strings attached. Why? Because it wants it that way. Nsyncerator was built not by a boy group but a bot group, made up of Eric Stoliker, Jim Snook, and Justin Bollman of Burbank, California.

The bot's primary weapon is a horizontal rotating blade. Its special features include an omnidirectional drivetrain. Nsyncerator took three months to build.

O

101

Just like the Secret Agent Man, they gave this robot a number, and took away its name. It responded by kicking a plethora of bot booty, winning its heat at the *Robot Wars 3* (UK) competition with consecutive victories over OverKill, Centurion, and King Buxton. In the semifinal heat, 101 won its opening match versus Scutter's Revenge but then failed in its attempt to make it into the tournament's final four when it lost to Hypnodisc.

At the *Robot Wars* 4 (UK) tourney, 101 came in seeded ninth. In the opening round it defeated Major Tom, then lost in the heat finals to Dominator 2.

ODIN II

A superheavyweight, Odin II split a pair of bouts during *BattleBots'* Season 3 tournament, winning its opener 28–17 over Whyachi, then losing in Round 2 to Electric Lunch 26–19. Odin II was built by John McKenzie of Ukiah, California.

OGRE

Built by Tom Griffin of Windom, Minnesota, Ogre made it to the third round of *BattleBots'* Season 3 superheavyweight tournament. Ogre defeated Flailbot 30–15 in the first round, outpointed Hoobot 34–11 in the second round, then lost a 23–22 decision in Round 3 to Minion.

OJ

Not named after orange juice, OJ was built by Victor Garaycochea of Boca Raton, Florida, and made it to the third round of *BattleBots'* Season 3 tournament. In its opener, OJ defeated Cyclone 30–15. It also outpointed Dreadnought 24–21 in Round 2 before losing a 23–22 decision to Phere in the third round.

OMEGA-13

Brent Regan of Coeur d'Alene, Idaho, created Omega-13, which did extremely well at *BattleBots'* Season 3 heavyweight tournament. In order, it defeated Scallywag 23–22, Bender 35–10, and B.O.B. 30–15, before losing to KillerHurtz in the fourth round by a 30–15 score.

ONE TIN SOLDIER

The loneliest bot in any arena is that solitary bot, One Tin Soldier. This wheeled and battery-powered bot features carbide prongs as its primary weapon. They are called Noogie Prongs by main builder Dane Scarborough of Hailey, Idaho.

Setting One Tin Soldier apart from the norm, many of its parts once belonged to a Ford Festiva. It took Scarborough and his helpers Hunter Scarborough, Joseph Lavigne, Richard Waligorski, and Michael Foley sixty days of planning and thirty days of building to complete One Tin Soldier.

At *BattleBots*' Season 2 superheavyweight competition, One Tin Soldier was eliminated in the opening round by War Machine.

ONSLAUGHT

At the *Robot Wars* 3 (UK) competition, Onslaught defeated Armour Geddon in the opening round of its heat, but was then defeated and eliminated in the second round by Beast of Bodmin.

ORION

Created by Christopher Hedberg of West Chicago, Illinois, Orion was KO'd in the opening round of *BattleBots*' Season 3 heavyweight tournament by Tripulta Raptor.

THE ORLANDO COMMANDO

Robert Wilburn and Team O-Town of Chuluota, Florida, constructed The Orlando Commando, which performed well in *BattleBots*' Season 3 middleweight tournament, chalking up a pair of KO victories before being eliminated. It KO'd Stealthbot—Tallahassee Lassee in the opening round, and Machiavelli in Round 2. It was subsequently defeated 34–11 by Garm.

OUTLAW

Outlaw is the brainchild of Mike Rede and Designated Destruction of Anaheim, California. It lost a 33–12 decision to Evil Con Carne in the opening round of *BattleBots*' Season 3 middleweight tournament.

OVERKILL

OverKill, which runs on battery power and wheels, has as its primary weapon a large blade made of the extremely tough heat-treated chrome-alloy steel. OverKill's body was made at Ballard and Sons CNC Machining.

In addition to attacking with its blade, OverKill also smashes and bashes its opponents like the best battering ram. OverKill's builder is Christian Carlberg of Santa Monica, California. His teammates on this project are Mike and Bernadine Ballard. OverKill is Toe Crusher's younger, but bigger, brother.

At the *Robot Wars* 3 (UK) competition, OverKill was eliminated in the opening round of its heat by the eventual heat winner, 101. In *BattleBots'* Season 2 heavyweight competition, OverKill was defeated by the heavy underdog, FrostBite.

At *BattleBots'* Season 3 tournament, Overkill defeated M.O.E. 28–17 in the opening round, KO'd RoboSapien in Round 2, outpointed Frenzy 23–22 in the third round, and defeated Voltronic 30–15 in Round 4. In the fifth round it was severely outpointed 38–7 by BioHazard.

OVERLORD

A superheavyweight built by Ronnie Pierce of Houston, Texas, Overlord lost a 32–13 decision to Cuad the Crusher in the first round of *BattleBots'* Season 3 tournament.

P

PACK RAPTORS

A tough and exciting bot, it is the creation of Chuck Pitzer and Team Raptor of Chandler, Arizona. Pack Raptors, Chuck's favorite bot out of all of his raptor bots, scored a pair of KO wins in *BattleBots'* Season 3 middleweight tournament before being eliminated. It zonked Squirrelly D and Gary Gizmo in the first two rounds, then was KO'd by T-Wrex.

PALINDROME

Palindrome, which looks the same comin' and goin', was KO'd in the opening round of *BattleBots'* Season 3 middleweight tournament by Cheap Shot. It was built by Anthony Hall and Team Radicus of Cockeysville, Maryland.

PANDORA

A veteran of the *Robotica* 2001 series on the Discovery Channel, Pandora, the bot, is named after the woman in Greek mythology who was given a jar by the God Zeus, which she was not to open under any circumstance. Impelled by her natural curiosity, Pandora opened the jar, releasing evil to spread over the earth. She closed the lid in time to save one thing: Hope.

Pandora the bot weighs 197 pounds, stands 24 inches tall, is 34 inches wide and 47½ inches long. It receives its power from two 12-volt Interstate 45ah deep-cycle gel-cell batteries. It has rack-and-pinion steering and a top speed of three miles per hour. Pandora's weapon is a steel flail.

PANIC ATTACK

Built by Kim Davies of England. At *Robot Wars* 3 (UK), Panic Attack won its opening heat, scoring consecutive victories over Axios, Toe Cutter, and X Terminator. In the semifinal heat, Panic Attack defeated Thing II in the opening round but failed to make it into the tournament's final four when it was eliminated by Fire Storm.

PANIC ATTACK 3

A heavyweight with wheels and battery power, Panic Attack 3 is another baby of builder Kim Davies of England. Davies's teammate is K. M. Pritchard. The bot's primary weapons are electric-powered lifting spikes. The bot has the ability to put itself right-side up in case it gets flipped over in combat and has the strength to lift twice its own weight.

PANZER

Named after the German battle tank, Panzer performed admirably in the United Kingdom. At the *Robot Wars* 3 (UK) competition, Panzer defeated Undertaker in the opening round of its heat, eliminated Challenger II in the second round, and then was defeated in the heat finals by Evil Weevil.

PARALLAX

Built by Jon Folkedahl and Team Attack of Willmar, Minnesota, Parallax lost in the opening round of *BattleBots*' Season 3 lightweight tournament by a 30–15 score to Sisyphus.

PARASITIC

Parasitic, from Jon Orrell and Team EC of Oregon City, Oregon, defeated Tarkus 27–18 in the opening round of *BattleBots*' Season 3 heavyweight tournament, then was KO'd in the second round by Incoming.

PATRIOT

Patriot is the work of George Roach and Team Bomb of The Woodlands, Texas. In *BattleBots'* Season 3 lightweight tournament, Patriot KO'd Silver Bullet in the opening round, then lost a 26–19 decision to Whirl Wep X25 in Round 2.

PHERE

Pronounced *fear,* naturally, Phere is a superheavyweight bot of some renown. In *BattleBots'* Season 3 tournament, Phere disposed of four opponents before it was bumped off. In the first round Phere KO'd Mantaray. In Round 2 it won a 25–20 decision over Go Postal. In Round 3 it defeated OJ 23–22, and followed that victory up with a 27–18 win over Anubis. In the fifth round, however, Phere was KO'd by Toro.

PHISHFEUXD

Pronounced *Phishfood,* this is a great jam bot, so put on your tie-dye and get ready to do the falling-leaf dance! The creation of Michael Macht and Team Osoka of Woodstock, Georgia, this bot was thrashed into lifelessness by Ravager in the opening round of *BattleBots'* Season 3 middleweight tournament.

PHOENIX

Not named after the city in the American Southwest, but rather after the mythological bird, Phoenix is the bot-child of Norman Muzzy and Team Flaming Monkeys of Cedar Falls, Iowa. Phoenix lost a 24–21 decision to WindChill in the first round of *BattleBots'* Season 3 middleweight tournament.

PHRIZBEE

Brian Nave and Team Logicom of Ormond Beach, Florida, created this whirling dervish, which lost badly to KillerB (40–5) in the opening round of *BattleBots'* Season 3 heavyweight tournament.

THE PIECEMAKER

A bit of clever wordplay from the right side of the pond, what? The Piecemaker was constructed by Mark Nenadic and a team he calls No

Mercy from Peterborough Cambs in the United Kingdom. This bot lost via KO to My Son in the opening round of *BattleBots*' Season 3 heavyweight tournament.

PINK SLIP

Not what a babe might slip into to get more comfortable, but rather what you find on your desk when you get fired, Pink Slip was created by Steven Doyle and Team Severance of Mountain View, California. It lost a decisive 40–5 judgment to Ice Cube in the first round of *BattleBots*' Season 3 middleweight tournament.

PIT BULL

Performed very well at the *Robot Wars* 3 (UK) tournament. Bred and trained for violence, Pit Bull won its heat with consecutive victories over Sump Thing, Technophobic, and B.E.H.emoth. In the semifinal heat, however, the vicious canine was defeated in the opening round by Firestorm.

PLUNDERBIRD 5

From England, built by the team of Mike Onslow and Bryan Kilburn, Plunderbird weighs 218 pounds and measures forty-seven inches by thirty-nine inches by sixteen inches. It runs on two 750-watt motors.

Plunderbird 5's strong point is that it is built very close to the floor, so it is difficult for wedgebots to get underneath it. Its weapon is a spinning blade.

In the "American Revolution: War of Independence" episode of *Robot Wars: Extreme Warriors,* in which bots from the United States were pitted against bots from the United Kingdom, Plunderbird 5, representing the right side of the pond, lost in its opening-round match to Rosie the Riveter. Plunderbird 5 made a good showing, however, knocking pieces off of Rosie with its initial contact and bending Rosie's spike downward into the floor so that it was all but incapacitated by the end of the fight. The match limped to a finish and Plunderbird 5 lost a close decision.

PLUNDERSTORM

At the *Robot Wars* 3 (UK) competition, Plunderstorm was eliminated in the opening round of its heat by Thermodor.

POKEY

Can Gumby be far behind? Pokey was built by Dallas Goecker and Team Sallad of Concord, California. It declawed Mechanical Wildcat in the opening round of *BattleBots*' Season 3 middleweight tournament, then lost a 35–10 decision to T-Wrex in Round 2.

PORTABLE KILLSAW

Sounds vicious, but went out like a lamb, KO'd by Huggy Bear in the first round of *BattleBots*' Season 3 middleweight tournament. Portable Killsaw was built by Robert Hamlet and Team BattleBobs of West Valley, Utah.

PRESSURE DROP

Pressure Drop is a walker, a stomp bot with a spiked hammer weapon. In the opening round of *BattleBots*' Season 1 middleweight tournament, Pressure Drop defeated Ankle Biter when Ankle Biter, who was winning the bout at the time, became wedged under the spikes along the walls. In the following round, Pressure Drop was taking a licking from a wedge named Subject to Change when the wedge lost power, allowing Pressure Drop to advance. In one of the more controversial moments in that tournament, Pressure Drop whacked Subject to Change with its spike after the bell. After two victories it didn't really deserve, Pressure Drop's luck ran out.

PRIZEPHITA

At the *Robot Wars* 3 (UK) competition, Prizephita was eliminated in the opening round of its heat by the eventual heat winner, Thing II.

THE PROBE

A favorite of alien abductors everywhere, The Probe was constructed by a team known, delicately enough, as The Penetrators. Head Penetrator is Jeff Banks of Napa, California. In *BattleBots*' Season 3 superheavyweight tournament, The Probe lost a 27–18 decision in the first round to Brainsters of Triskellion.

PROMPT CRITICAL

Promp Critical broke down during its opening match in *BattleBots'* Season 1 superheavyweight tournament, and was eliminated by that contest's Medal of Honor winner, the spiderlike Mechadon.

PSYCHO KILLER

You better run, run, run, run, run, run away. Or, maybe not. Psycho Killer didn't excel in its only fair fight, being defeated by Dreadnought in the opening round of its heat at the *Robot Wars* 3 (UK) tournament.

PSYCHO SPLATTER

This tastefully named bot hails from Anoka, Minnesota, where it was built by Joel Remer and a team (also tastefully named) known as Flying Parts. Psycho Splatter was itself dismembered in the opening round of *BattleBots'* Season 3 heavyweight tournament by Bad Badger.

PSYCHOTRON

Kelly Smith and Team Rockitz of San Dimas, California, built Psychotron, which lost a 23–22 decision to Thwak in the first round of *BattleBots'* Season 3 middleweight tournament.

PUNJAR

Punjar is a wheeled and battery-powered heavyweight built by Ramiro Mallari of San Francisco, California. The wedge-shaped bot uses high-speed attacks to dish it out and tough armor to take it. Punjar, according to Mallari, took three months to build. Punjar was knocked out in the opening round of *BattleBots'* Season 3 heavyweight tournament by Shark Byte.

PURPLE HAZE

Uh, excuse me while I kiss the sky. There. OK, back to business. This bot was all in the brain of Nick Belton. It is the Alfred University (Alfred, New York) entry. At *BattleBots'* Season 3 lightweight tournament, Purple Haze realized that it was not tomorrow, but just the end of time, as it was KO'd by Bulldog in the opening round.

PURPLE PREDATOR

You've heard of Purple Passion, right? Thus, Purple Predator. At the *Robot Wars* 3 (UK) tournament, Purple Predator was eliminated in the first round of its heat by Toe Cutter.

PURSUER

Darren Mackenzie of St. Petersburg, Florida, built Pursuer. At *Battle-Bots'* Season 3 lightweight tournament, Pursuer advanced past the first round bcause of a forfeit, then lost a 27–18 decision to Sisyphus in Round 2.

PUSSYCAT

Pussycat can't keep its claws in. At the *Robot Wars* 3 (UK) competition, this feline fighter scratched its way to victory over Hammerhead in the opening round of its heat, then followed that victory up with another win over Cassius II before losing by disqualification in the heat finals to Scutter's Revenge.

At *Robot Wars* 4 (UK), Pussycat won its heat with victories over Robochicken and Razer.

In the "American Revolution: War of Independence" episode of *Robot Wars: Extreme Warriors,* in which bots from the United States were pitted against bots from the United Kingdom, Pussycat, representing the British Empire, KO'd the Six Million Dollar Mouse in the opening round, using its spinning blade to cut the nose clean off the rodent. In the semifinals, Pussycat overcame a big size and weight disadvantage to defeat Rosie the Riveter. The climax of that bout came when both bots fell into "The Pit," but, because Rosie went down first, Pussycat was awarded the victory. In the tournament finals, Pussycat defeated General Chompsalot to rewrite history and vanquish the revolution. In that final match, Pussycat's blade chewed up General Chompsalot until the big-mouth Yank bot was bearly recognizable.

Pussycat was built by Alan Gribble, David Gribble, Robert Bettington, and Team Cold Fusion. Alan's advice to potential robot builders: "Be prepared to do all of the work yourself. No one will be as committed as you are. I admire anyone who actually gets up off their butt and manages to build a bot and make it to the auditions."

Alan predicts that in five years combating robots will be dominated by big-business sponsors and will be more like the World Wrestling Federation or Formula 1 Grand Prix racing. Small-time builders, like himself, he says, will be lost to history.

PYRAMIDROID

Built by Mike Konshak of Louisville, Colorado, PyRAMidroid lost by a score of 32–13 in the opening round of *BattleBots*' Season 3 heavyweight tournament to Bacchus.

R

RA

Named after the favorite saying of a truly concise cheerleader, the bot with the shortest monicker is Ra, which was built in approximately two weeks by Solomon Specter of Goodlettesville, Tennessee.

It is built on wheels and powered by batteries. Its primary weapon is a movable kill-saw tail. In the front it has two kill-saw claws. The bot's hard shell is composed of Kevlar 29 reinforced concrete.

RAIZER BLADE

Raizer Blade defeated Red Dragon in the opening round of *Robot Wars* 3 (UK) tournament, then was defeated and eliminated in the second round by Suicidal Tendencies.

RAM FORCE

"The Chairman of the Circuit Board." Whenever they get around to creating a bot hall of fame, the 208-pound Ram Force is a lock to be one of the first honored in those hallowed halls.

Built by Andrew Lindsey, age twenty-eight, of Montclair, New Jersey, and Tony Altenoeder, of Eden, New York, Ram Force features compact, high-density nickle-cadmium batteries, custom-made electronic control gear, high-precision industrial 2-horsepower robotic servo drive motors, bulletproof planetary gearboxes, massive grippy racing tires, and a precision-machined aluminum frame.

Lindsey says that the most important part of his bot is traction and control. It is also very difficult to injure because of its bulletproof 1¼-inch Lexan armor. Because of its modular design—pieces can be removed and replaced with great ease—Ram Force can be modified at a moment's notice to best match up against a foe. Though this one is a classic, Lindsey says that it is merely a prototype for future, better designs.

Team Force is a veteran of many bot competitions, including Atlanta DragonCon 1996, Robot Wars 1996 (semifinalist), Robot Wars 1997 (semifinalist), BotBash 1998 (first place), BattleBots 1999, BotBash 2000 (third place), Robot Wars UK 2000. Lindsey also served at BattleBots Las Vegas 2000 as a combat judge.

Ram Force was the winner in its first appearance on *Robotica,* during Season 1, defeating HandsOff! in the preliminary rounds and finishing off Sword Player in the Fight to the Finish. This allowed Ram Force to advance to the Final Six.

In that round, it met Jawbreaker in combat. Ram Force won on the Speedway and in the Maze but lost in the Gauntlet. This was still good enough to earn it a 250–160 win and advancement to the three-way final Fight to the Finish, where it was eliminated by Run Amok.

One last word of wisdom from Andrew Lindsey: "Never pass up an opportunity to mess up the other robot."

RAMMING SPEED

The name of the bot and its method of operation are one and the same. The brainchild of Joseph Fieber and Team Clobber of St. Francis, Wisconsin, Ramming Speed lost a 25–20 decision to The Archduke in the opening round of *BattleBots'* Season 3 lightweight tournament.

RAMMSTEIN

Korey Kline of Miami, Florida, is the main builder of Rammstein, a wheeled and battery-powered superheavyweight that uses a high-pressure (800 pounds per square inch) pneumatic spike as its primary weapon. Rammstein has six-wheel drive and functions equally well when upside down or right-side up. The bot features a shell of $\frac{3}{16}$-inch aluminum plate in welded monocoque construction.

Kline's teammates on the project are Alex Espinosa, Eddy Ampuero, Julio Roqueta, Mercy Cruz, and Nola Garcia.

At *BattleBots'* Season 1 superheavyweight tournament, Rammstein survived as far as the semifinals, where it was eliminated by the eventual winner, Minion.

In Season 2, Rammstein advanced as far as the second round before being eliminated by the eventual winner, Diesector.

In the Season 3 tournament, it KO'd its first two opponents, Half Gassed and The Judge, before being KO'd itself in Round 3 by Diesector.

RAMPAGE

A solid-looking bot that competed in Season 2 of *BattleBots,* and lost in the first round to Backlash. Rampage featured steel armor, a spinning blade, and three menacing spikes. In a bout that did not air on television, Rampage's side spike was torn off by Backlash. The spike flew all the way across the arena and ricocheted off two of the walls before crashing back to the floor.

Rampage was built in about 250 hours of labor by Troy Racey of Baltimore, Maryland. Battery powered and built on wheels, Rampage's primary weapon is a 5¼-inch vertical blade mounted to the front. In the rear, Rampage sports three stationary spikes, which it uses predominantly to fend off attacks.

RATTUS RATTUS

Named by an echolaliac, Rattus Rattus defeated Smidsy in the opening round of its heat at the *Robot Wars* 3 (UK) tournament. It was then eliminated in the second round by Dreadnought.

RAVAGER

Ravager made it to the third round of *BattleBots'* Season 3 middleweight tournament, knocking out Phishfeuxd in its opener, and winning a 25–20 decision over No Tolerance II in Round 2 before being eliminated via KO by Sunshine Lollibot.

Built by Eston Logston of Redding, California.

RAZER

Razer was built by Vinny Blood, who says, "We know that Razer is capable of beating just about any robot out there. So long as Lady Luck smiles on us we are always in with a chance. This may sound arrogant to some, but if you are to win you need to have confidence in yourselves and your machine."

At the *Robot Wars* Series 2 (UK), which was taped for broadcast in the United Kingdom in August 1998, Razer finished first in the Gauntlet competition, took first place in the football (soccer) game, but lost in the finals to Inquisitor after its nylon cush drive was sheared. Razer did not go home empty-handed from the tournament, however, as it was the winner of the Best Design award.

At the Los Angeles *BattleBots* tapings of August 1999, it defeated Voltarc by decision, knocked out Aggrippa, and then lost by KO to Tazbot.

At the *Robot Wars* 3 (UK) tourney of October 1999, Razer KO'd Backstabber but lost to Aggrobot in the second round after a hydraulic valve jammed. At that taping, Razer was again the winner of a Best Design award, and was also the winner of the "Pinball Warrior" tournament.

At the Robot Wars First World Championship of October 1999, Razer KO'd All Torque in the opening round, defeated Chaos II in the second round (after Chaos II drove into the pit), defeated 101 in Round 3 by decision, and then outpointed B.E.H.emoth in the finals to win the tournament.

In the Robot Wars International League competition, also held in October 1999, Razer defeated Prometheus when Prometheus retired, then KO'd Technolease and Diotoir to win the tournament.

At the *BattleBots* taping held in Las Vegas in November 1999, Razer made an uncharacteristic exit in the opening round, losing by decision to Voltarc.

At the *Robot Wars* 4 (UK) tourney, Razer looked sharp as it won a decision over Robochicken, KO'd Milly Ann Bug, then lost in a close shave to Pussycat in the heat final. The loss came after Razer's fail-safe switch failed.

REACTORE

Reactore was created by Michael Bastoni and the Pilgrim Nuclear Technology Alliance of Plymouth, Massachusetts. In the opening round of *BattleBots*' Season 3 heavyweight tournament, Reactore was KO'd by Botknocker 19.

REAPER

It's amazing what you can get done if you never sleep. Just ask Robert Wooden and Team Caffeine of Burlington, Wisconsin, the proud parents of Reaper, which won a 34–11 decision over Daisy in the opening round of

BattleBots' Season 3 lightweight tournament. Reaper then KO'd Wedge-o-Matic in the second round before losing a 23–22 squeaker to Bad Habit in Round 3.

REBOB

Constructed by Jason Woods and New Tech Robotics of Napa, California, Rebob was KO'd by Eraser in the opening round of *BattleBots*' Season 3 middleweight tournament.

RECYCLER

Don Jennings and Team Recycle of Omak, Washington, made the most of stuff they had lying around when they built Recycler. After advancing past the opening round of *BattleBots*' Season 3 lightweight tournament because of a forfeit, Recycler won a 34–11 decision over Brain Mold in Round 2, then was KO'd in the third round by Whirligig.

RECYCLOPS

Recyclops was the winner of its heat at the first *Robot Wars* UK series, but lost in the finals to Roadblock.

RED DRAGON

Red Dragon was eliminated from the *Robot Wars* 3 (UK) tournament in the opening round by Raizer Blade.

RED SCORPION

Red Scorpion is a stompbot—that is, it walks on legs rather than rolls on wheels. It does not function on battery power like so many of today's battling robots, but rather operates on compressed carbon dioxide.

It was built in "hundreds of hours" by Shane, Jon, and Ray Washburn, along with Andy Miller, all of San Leandro, California.

REPTIRRON

At the *Robot Wars* 4 (UK) tourney, Reptirron lost in the opening round of its heat to Milly Ann Bug.

REVISION Z

Revision Z is a remodeling of the old robot known as Grendel. It runs on wheels and battery power and has a linear-actuator-driven lifting arm as its primary weapon. The bot, in its latest configuration, took four months to build. The main builder is Daniel Rupert, who completed the project with the aid of teammates Scott Robeson, Steven Schulze, and Seth Herring, all of San Diego, California.

At *BattleBots*' Season 2 superheavyweight tournament, Revision Z won a shoving match against Gray Matter in the opening round, then defeated Ronin in the second round when Ronin caught on fire. In the semifinals, Revision Z was eliminated by Atomic Wedgie.

At *BattleBots*' Season 3 tournament, Revision Z KO'd Gray Matter in the first round and then lost a 30–15 decision to Vladiator in Round 2. (See also GRENDEL.)

THE REVOLUTIONIST

Built by Brian, Michael, and Rebecca Nave from Florida, The Revolutionist is a round spinner that weighs 207 pounds and measures twenty-eight inches by twenty-eight inches by seven inches. It runs on four electric motors. The Revolutionist's weapons are its revolving cutting teeth.

In the "Civil War" episode during the first season of *Robot Wars: Extreme Warriors,* The Revolutionist represented the South and lost a close decision in the opening round to Sobek.

RHODE HOG

A pesky porcine bot, Rhode Hog was built by Michael Thombs and Team Clam Bot of Portsmouth, Rhode Island. In *BattleBots*' Season 3 lightweight tournament, Rhode Hog won a 30–15 decision over Archangel in the opening round, advanced past Round 2 when its opponent, Road Dot, was unable to compete, and then lost via KO to Matts Bammer in Round 3.

RIM TIN TIN

Crossing robotics with man's best friend is as old as the hills. Just look at Dr. Who and his K9 buddy. Rim Tin Tin was best friends with Dan Watts, Christina Zaccherio, and Team Solo of New Hyde Park, New York.

It took approximately six weeks to build. It is battery powered, runs on wheels, and its weapons are a plow blade and a spike. Rim Tin Tin was built out of a steel car rim discovered by Watts when he was employed as a tow-truck driver.

At *BattleBots*' Season 3 lightweight tournament, Rim Tin Tin was sent to doggie heaven in the opening round by HammerHead.

THE RIPPER

The Ripper's daddy is Brian Ondov of Whitehouse Station, New Jersey, who created his bot child with the help of Team Servo. This bot lost 24–21 to Icky Thump in the opening round of *BattleBots*' Season 3 heavyweight tournament.

RIPOFF

RipOff isn't named after stealing, it's named after "steeling," as its dual vertical spinning twenty-eight-inch blades will rip parts right off of any competitor. The wheeled and battery-powered superheavyweight was built in two months by main builder Christopher Hannold of Linwood, North Carolina, with the help of teammates Chris Crowell, Angela Crowell, and Greg Hannold. How did Hannold get the idea for this excellent weapon system? He stole it, that's how. Thus the name of the bot. (Oh, OK, it *is* named after stealing.)

ROADBLOCK

Back in 1997, builder Hender Blewitt was a tech student in Cornwall, England, looking for a project. As if by an act of God, just as this was occurring, Hender was given information detailing the upcoming *Robot Wars* TV series.

Along with a teacher—Peter Kinsey, Head of Technology Department at Bodmin Community College—and fellow student Chris Kinsey, Hender set out to build the baddest bot in the world.

The result was Roadblock. The name had originally been Roadrage, but was toned down at the request of the TV show. At that competition, Roadblock proved to have superior power and torque. Roadblock was the winner of the 1998 *Robot Wars* series 1 (UK).

The chassis of the robot is based around a 24-volt electric wheelchair motor running off two 12-volt gel-cell batteries. The original control box was modified to suit the application, providing a versatile method of control.

Its weapons are its wedge front and a circular saw in the rear. The blade is powered by a 25-cubic-centimeter two-stroke strimmer engine. The blade is capable of spinning at 6,600 revolutions per minute.

According to Blewitt, "Initial tests proved that hardened steel blades were more manly than the carborundum equivalent, so we obviously opted for steel immediately. The robot was built within three months of the application and was running after two and a half."

The bot turned out to be so powerful that it could pull Hender's car around a parking lot. The thing that people notice first about Roadblock is that its shell is constructed of road signs that were "donated."

Roadblock weighs 82½ kilograms. It measures 145 centimeters by 110 centimeters by 40 centimeters. It has a ground clearance of 1 centimeter and a top speed of five miles per hour.

ROBOCHICKEN

At the *Robot Wars* 4 (UK) tourney, Robochicken defeated Velocirippa in the opening round of its heat, but was eliminated in the second round by Pussycat.

ROBO COW

This bovine bot lost in the opening round of the *Robot Wars* 3 (UK) tournament to General Carnage.

ROBOGEDDEN

Robogedden was eliminated in the opening round of its heat at the *Robot Wars* 3 (UK) competition by the eventual heat-winner, Hypnodisc.

ROBO MASTER

Robo Master may have been master of its own domain (especially since it lacked arms), but it mastered little else, as it lost a crushing 41–4 decision to Matts Bammer in the opening round of *BattleBots*' Season 3 lightweight tournament.

Robo Master was built by Jim Yeh of Piedmond, California.

ROBO PIG

This porcine pug defeated Bulldog Breed in the opening round of the *Robot Wars* 3 (UK) competition, then was eliminated in the second round by Napalm.

ROBOSAPIEN

A fearsome competitor, RoboSapien was built by Robert Hamlet and Team BattleBobs of West Valley, Utah. This bot KO'd Jolly Roger in the opening round of *BattleBots*' Season 3 heavyweight tournament, then won a lopsided 42–3 decision over Bad Badger in Round 2, then lost to OverKill via KO in the third round.

ROBOT THE BRUCE

Robot the Bruce was the winner of its heat at the first *Robot Wars* UK series, but lost in the finals to Roadblock.

ROBOT X

Edward Robinson and Team Blaze of Muskogee, Oklahoma, built Robot X, which lost a brutal 41–4 decision to Fang in the opening round of *BattleBots*' Season 3 lightweight tournament.

RONIN

Ronin is the brainchild of Peter Abrahamson of Burbank, California, who built it with the help of teammates Eric Abrahamson, Cassidy Abrahamson, Roe Leone, and Terry Sawdin.

The bot is battery driven and runs on wheels, but its primary weapon is a gas-powered twenty-inch-diameter saw. That saw can be quickly replaced with a battery-powered lifting sword, if that is deemed the more effective weapon versus a given opponent.

According to Abrahamson, "Ronin's tread pods are independent and can pivot on the main two-inch-diameter support shaft. My bot has modularity. I can change my weapons to better suit the battle with the other robot." Abrahamson says the bot took "thousands of hours" to build. Ronin has been around for a while and has undergone many changes and metamorphoses.

At *BattleBots'* Season 2 superheavyweight competition, Ronin was set afire and eliminated in the second round by Revision Z.

In *BattleBots'* Season 3 tournament, Ronin lost by KO in the opening round to Electric Lunch. Back to the drawing board.

At least Abrahamson has got his priorities straight: "There are about four things I cherish in life," he says. "Robotic combat, rock climbing, fly fishing, and sex—not necessarily in that order."

ROSIE THE RIVETER

Rosie was built by Oregon's Chris Gattman, Sharon Gattman, and Kris Ender of Team Thunderbolt. The bot weighs 216 pounds and measures forty-three inches by thirty-six inches by twelve inches.

It is powered by a quartet of wheelchair motors. Its primary weapon is its iron battering ram, which has not proven to be effective in combat. The bot's strong point is its speed.

Rosie competed in the Season 1 Tag Team Terror Tournament, teamed up with Tricerabot. They lost their opening match to Tut Tut and Drillzilla, but emerged victorious in the battle of the losers that followed, defeating Joker and Falcon.

In the "Robot Rebellion" episode of *Robot Wars: Extreme Warriors,* in which entrants competed for the right to battle *Robot Wars'* house bots, Rosie defeated Rot Box in the eliminator round, earning the right to be one of four winners to take on three house bots—Shunt, Sir Kill-a-lot, and Matilda—battle-royale-style, a melee that the larger and more impressive house bots unexpectedly lost.

In the "American Revolution: War of Independence" episode of *Robot Wars: Extreme Warriors,* in which bots from the United States were pitted against bots from the United Kingdom, Rosie defeated England's Plunderbird 5 in the opening round. Plunderbird 5's first contact knocked pieces from Rosie but Rosie's superior ramming power got it back into the fight. With a decision that could go either way, Rosie had her spike bent downward so that it dug into the floor and rendered the bot practically motionless. Luckily, Plunderbird 5 was in slightly worse shape and Rosie won by a close decision. In the semifinal bout, Rosie had a big size and weight advantage over the eventual tournament winner, Pussycat. During the early stages of the fight, Rosie was pushing Pussycat all over the dance floor. But Pussycat came back and did great damage to Rosie with its spinning blade before both bots fell into "The Pit." Since Rosie went into the hole first, Pussycat was awarded the victory and advanced on to the finals.

ROT BOX

Built out of wood by a pair of farmers named Jason Berger and Brent Hunter of Pembroke Pines, Florida, Rot Box weighs 214 pounds and measures thirty-nine inches by thirty-one inches by twenty-four inches.

It is powered by two electric motors and its primary weapon is its pneumatic mace. It is a strong bot but at an obvious disadvantage because of its wood construction. In the "Robot Rebellion" episode of *Robot Wars: Extreme Warriors,* in which entrants competed for the right to fight the *Robot Wars*' house bots, Rot Box was knocked out in the eliminator round by Rosie the Riveter.

ROTT-BOT 2000

Rott-Bot 2000 was eliminated in an early round of *BattleBots*' Season 1 lightweight tournament by Das Bot. Rott-Bot 2000 made the highlight reel during that bout when it accidentally meandered onto a kill saw and temporarily became an antigravity bot.

RUN AMOK

Run Amok is the creation of Mark Joerger, a forty-nine-year-old researcher for the Oregon Department of Transportation, who also performs as a magician at parties, and his ten-year-old son, Aaron Max.

If this bot seems to have an international feel, it's because its parts came from here, there, and everywhere. The chassis started out inside a good old-fashioned American riding lawn mower, but the electric motor came from Germany, the speed controller was brought in from England, and the radio gear and steering servo come from Japan.

Run Amok was the surprise winner of *Robotica*'s Season 1 tournament. In its first appearance on the show, Run Amok was matched up against Spring Breaker in the preliminary rounds.

It won that round and then defeated KillerB in the Fight to the Finish. This advanced Run Amok to the Final Six, where it was matched up against Kritical Mass in the preliminary rounds.

In the Speedway event, Run Amok completed six laps to Kritical Mass's five. In the Maze, Kritical Mass made a comeback, getting past five obstacles to Run Amok's two. In the Gauntlet, Kritical Mass stalled and Run Amok burned out its engine to push past one final obstacle. But it did push past, and this enabled Run Amok to win the Gauntlet. The score was 210–210, but since Run Amok had won two of the three events it advanced to the three-way final Fight to the Finish against JuggerBot and Ram Force.

The team that put JuggerBot together actually helped Run Amok fix the engine between rounds, as they viewed Ram Force as the true threat to their dominance. In the finals, Ram Force and JuggerBot locked horns and pushed with all of their might while Run Amok bided its time, occasionally dishing out a halfhearted blow here and there. When the guardrails came down, Run Amok changed its attitude and cleaned house for the victory.

Mark Joerger describes how he first became aware of the sport that has become his almost-obsession: "I started seeing little news clips and snips of video back around 1995 talking about battling robot demonstrations and contests in the San Francisco Bay Area. I kept my ears open and followed along as best I could, but info was sparse. I immediately recognized that this was something that I would like to try out and that it was a good match for my skills and experience. I had a history of competing in odd mechanical contests: manned and unmanned gravity racers, assorted miniature hot air balloons, various remote-controlled vehicles and electric racers. It took a few years to connect with the competitions and start building. I'm not sure that it has ever developed into an obsession, but it does absorb a lot of my time."

One thing that troubles Mark about the development of robotic combat is the skyrocketing costs. "The amount of money and time required to be competitive seems to be rising at an alarming rate!" he says. "It used to be possible to build a nice little $2,000 bot and be competitive, but not anymore. I guess it's like most types of competition—it's hard to beat cubic money."

Indeed, Mark feels that corporate sponsorship is already a necessity to compete. "You already need deep pockets or major sponsorships to be in the top echelon, although spending great wads of cash certainly won't guarantee success," he says.

His other pet peeve: "There is also a push by event organizers to require or strongly favor 'active weapons' on the bots. I suppose the organizers believe that it's more entertaining to watch bots with moving weaponry, but I like to root for the wedges and pushbots."

RUN AWAY

Run Away was, like Run Amok, built by Mark, Aaron, and Lissa Joerger of Salem, Oregon. This wedge-shaped bot weighs 218 pounds and measures fifty-six inches by thirty-five inches by thirty-one inches. Run Away is powered by two 24-volt electric motors. Its weapons include both a scoop, designed to get under the opponent and flip it over, and a pair of spinning blades. Run Away participated in the Tag Team Terror Tournament during Season 1 of *Robot Wars: Extreme Warriors.* In that competition, Run Away teamed up with General Chompsalot. That team defeated the team of Joker and Falcon, but lost in the finals to Tut Tut and Drillzilla. Run Away is suspiciously similar to Run Amok, which competed on the *Robotica* TV series. They are different bots, however. Says Joerger: "Run Amok was purposely built for the unique challenges at the first *Robotica* competition—high clearance, waterproof top armor, plenty of speed. It proved very successful at what it was designed to do, considering it started its robotic life as a discarded lawn tractor. When we were offered a chance to compete at *Robot Wars,* I took a look at Run Amok and decided it would be an interesting and entertaining *Robot Wars* competitor if I added a big visible weapon and made it a little more rugged and tip-proof. The twin-rotating-blade weapon was fabricated, bolted onto an enlarged top armor panel, and patched into the remote-control system via a power relay from an electric car. The front steering system was redesigned to absorb more punishment, and additional fail-safe electronics were incorporated. Run Away was born."

What are Mark Joerger's dreams for the future of the sport?

Mark says, "I would like to see the focus of the sport expand beyond simple destructive combat. The diverse challenges of the *Robotica* competition are a good start in that direction. Challenges that reward creativity and innovation could take some of the emphasis away from the high-dollar super-bots. Also, new competitors should be aware of the many regional and local robot combat events staged across the country. These events don't get TV coverage, but they are a lot of fun. They are also much less expensive to get started with, as they have weight classes for bots as small as one pound. You can gain a lot of experience and knowledge inexpensively by competing in these events!"

S

SABOTAGE

Matt Sabotino of Greenwood, Indiana, named this bot after himself. Actually, much of the Sabotino family was involved in Sabotage's creation, with Michael and Susan lending a hand.

Sabotage defeated Huggy Bear in the opening round of *BattleBots'* Season 3 middleweight tournament 23–22, won a 34–11 decision over Double Agent in the second round, then KO'd Twin Paradox in Round 3. In Round 4 it was eliminated by Little Drummer Boy.

The wedge-shaped Sabotage runs on a 24-volt battery, but its weapon—a lifting and throwing arm—is fueled by compressed carbon dioxide. Other features include dual drive, a chrome moly frame, Lexan armor, and a Vantec speed controller.

SALLAD

Built by Dallas and Mandy Goecker of Concord, California, Sallad was built very low to the ground. It has a nasty arm that triples as a lifter, a hammer, and a ramming spike. That arm can swing greater than 180 degrees.

Sallad operates with either side up. It was eliminated in the opening round of *BattleBots'* Season 1 lightweight tournament by a bug-shaped bot named Tentoumushi. Sallad was tossed in just under two minutes by Backlash in the quarterfinals of *Battlebots'* Season 2 tournament.

At the Season 3 tourney, Sallad KO'd Carnage Raptor and Sublime in the first two rounds. It then lost a 36–9 decision to the eventual winner of the tournament, Dr. Inferno Jr. Sallad lost all chance to win the round when it became stuck under the spikes that lined the BattleBox wall.

SCALLYWAG

I would have named this bot Killer Wales, but no one asked me. Scallywag was created by Kim Davies and the Welsh Wizards, who are from Cwmbran Gwent, Wales, United Kingdom.

In *BattleBots'* Season 3 heavyweight tournament, Scallywag lost in the opening round in a squeaker, losing 23–22 to Omega-13.

SCAR

At the *Robot Wars* 4 (UK) tourney, Scar lost in the opening round to The Morgue.

SCARAB

Scarab weighs 210 pounds and was built of steel bar stock and Cyrolon acrylic. It features extremely high torque and traction. It can turn very tightly and has high clearance, all attributes that make it perfect for *Robotica,* where it last competed in 2001.

Scarab features a 220-amp, 24-volt drive system with four 1.6-horsepower NPC motors driven off two 12-volt lead-acid batteries throttled through a Vantec controller. Scarab can use its pneumatic system to lift over 1,000 pounds in one quarter of a second.

The controller combines off-the-shelf components for the drive system and custom-built electronics for the pneumatics. Scarab also has a nifty camera for getting a bot's-eye view through a remote heads-up display.

It was built by Martin Calsyn, age forty-two, Bronwyn Calsyn, age thirty, David Buckley, age thirty-one, and Evan Marshall, age fourteen, of Carnation, Washington. Martin and David are software design engineers. Bronwyn is a systems analyst and Evan is a student.

At the *Robot Wars* 3 (UK) tournament, Scarab was eliminated in the opening round of its heat by Victor II.

SCHEDULED FOR DESTRUCTION

Built by John Hoffman and Team Kirwan of Poway, California, Scheduled for Destruction knocked out El Cucuy in the opening round of *BattleBots'* Season 3 middleweight tournament, then lost a 24–21 decision to Double Agent in the second round.

SCHUMEY II

At the *Robot Wars* 3 (UK) tournament, Schumey II lost in the opening round of its heat to Steel Avenger.

SCRAP DADDY HW 210

Scrap Daddy HW 210, a heavyweight bot on wheels, was built in "a few weeks" by Mark Bradford, Sean Ham, Dick Craig, and Elizabeth Winningham of Houston, Texas. Although it runs on battery power, its blade weapon is powered by liquid fuel.

SCRAP DADDY LW55

The lightweight entry from Mark Bradford and Elizabeth Winningham of Houston, Texas. Built on wheels and operated by electricity, Scrap Daddy LW55 has a blade weapon that is powered by liquid fuel. According to Bradford it took one week to build.

SCRAP DADDY MW110

Built by Mark Bradford, Sean E. Ham, and Elizabeth Winningham of Houston, Texas, this middleweight Scrap Daddy is, like the others, on wheels and battery powered. It was built in "a few weeks," and has a blade weapon that operates on liquid fuel.

SCRAP DADDY SHW325

The biggest of all the Scrap Daddies weighs 325 pounds. This super-heavyweight, built by Mark Bradford of Houston, Texas, with the help of teammates B. J. Walter, Dick Craig, Elizabeth Winningham, and Ken Crimins) on wheels, is battery powered, and has a blade weapon that operates on liquid fuel.

SCRAP DADDY'S FLIPSKANKER

Scrap Daddy Mark Bradford of Houston, Texas, built this middleweight, which was knocked out by T-Wrex in the opening round of *BattleBots'* Season 3 tournament.

SCRAP DADDY'S PERSISTOR

As those above, this bot was built by Scrap Daddy himself, Mark Bradford of Houston, Texas. Scrap Daddy's Persistor lost to Golddigger by a score of 30–15 in the first round of *BattleBots'* Season 3 heavyweight tournament.

SCRAP DADDY'S SURPLUS

Mark Bradford's entry in *BattleBots'* Season 3 lightweight tournament, Surplus lost a 30–15 decision in the opening round to Herr Gepoünden.

SCRAP DADDY'S TURBULANCE

Another bot by Mark Bradford of Houston, Texas. Scrap Daddy's Turbulance lost in the opening round of *BattleBots*' Season 3 superheavyweight tournament, knocked out by Diesector.

SCRAP METAL

Scrap Metal, a lightweight veteran of *BattleBots,* was built in 200 hours by Robert Masek, Chris Wolfson, and Alan Kelley of Nashua, New Hampshire. Built on wheels and battery powered, Scrap Metal's weapons are kinetic energy, a rotating body, and its wedge shape. It has modular construction and a strong drivetrain designed for a heavyweight bot.

SCRUB

A lightweight bot built in two days by Mike and Doris Regan of Redondo Beach, California. It is a smaller version of Spaz, Regan's middleweight bot. This ruggedly built and battery-powered bot has a flailing spiked tail as its main weapon.

In *BattleBots*' Season 2 tournament, Scrub was eliminated when it lost a 43–2 decision, in a match that did not air on television, to the eventual second-place finisher, Backlash.

SCUTTER'S REVENGE

At the *Robot Wars* 3 (UK) competition, Scutter's Revenge won its heat with consecutive victories over Zeus, Thermodor, and Pussycat (by disqualification). In the semifinal heat, Scutter's Revenge lost in the opening round to 101.

SGT. BASH

Sarge is a house bot on the *Robot Wars: Extreme Warriors* TV show. It weighs 264 pounds and measures fifty-six inches by thirty-six inches by thirty-six inches. It is powered by four parallel batteries and has two main weapons. First is a set of pincers that can grab a foe so that they know they've been grabbed, and a flamethrower. The flamethrower makes Sgt. Bash far more dangerous than most bots because it can attack from great distances, whereas most bots actually have to make contact with another bot before they can dish out the damage.

SGT. MEIKLE

At the *Robot Wars* 3 (UK) competition, Sgt. Meikle lost to Darke Destroyer in the opening round of its heat.

SGT. SPANKY

This bot was built by Patrick Urschel of Valparaiso, Indiana. Sgt. Spanky was KO'd in the opening round of *BattleBots*' Season 3 heavyweight tournament by Incoming.

SERIAL BOX KILLER

Built by Matthew Garten, Justin Garten, Ryan Foss, and Team Minimal Control of Goleta, California, Serial Box Killer is built on wheels and runs on battery power.

Serial Box Killer is a rammer and can operate equally well no matter which side is up. It has ¼-inch-thick aluminum armor, which gives it the strength of Tony the Tiger.

Serial Box Killer lost to Backlash in the opening round of *BattleBots*' Season 1 lightweight tournament.

It won a 24–21 decision over Junkyard Dog in the opening round of *BattleBots*' Season 3 tournament, outpointed Manic Aggressive 30–15 in Round 2, then was KO'd in the third round by Gamma Raptor.

SHADOW OF NAPALM

At the *Robot Wars* 4 (UK) tourney, Shadow of Napalm came in seeded twenty-fifth, but lost in the opening round of its heat to the unseeded (and eventual heat winner) Dominator 2.

SHAFT

Shaft was built by Rik Winter and the Robot Action League of Oakland, California. Shaft lost an opening-round decision to Mouser Mecha-Catbot at *BattleBots*' Season 1 lightweight tournament. It was a 26–19 loser to Blood Moon in the opening round of the Season 3 tournament.

SHAKA

Shaka was constructed by Walter Martinez and Team DreamDroid of Lakewood, California. It lost a 29–16 decision to the eventual tournament winner, Son of Whyachi, in the opening round of *BattleBots*' Season 3 heavyweight tournament.

SHANNONAGAINS

Shannonagains was modified, or I should say customized, from a Power Wheels toy truck. It is now way souped up, with four-wheel drive, 24-volt power plant, power angle plow, a forklift, a retractable spinning mass weapon, AM/FM stereo cassette player, and working lights.

It can be driven both with a radio control or by an actual driver. After its modifications, the Power Wheels truck weighs in at 206 pounds. It was built by Thomas Rogers, age thirty-nine, a carpenter from Lake Peekskill, New York.

He says that his favorite Hollywood bot is Data from *Star Trek: The Next Generation.* His hobbies include music and antique cars. His teammates include Shannon Rogers, after whom the bot was named, and Debra Pierce. Shannonagains made its TV debut at *Robotica II.*

SHARK ATTACK

Shark Attack lost in the opening round of the *Robot Wars* 3 (UK) tournament to B.E.H.emoth.

SHARK BYTE

They called 2001 the year of the shark, and that almost became the case in the world of battling robots as well. A force to be reckoned with (and feared), Shark Byte was built by Arndt "The Enforcer" Anderson and Team Shark of Ione, Washington.

In *BattleBots*' Season 3 heavyweight tournament, Shark Byte advanced past Spinning Mayhem in the opening round, sent Heart of Darkness into cardiac arrest in Round 2, hermetically sealed Punjar in the third round, and then lost by KO to Hexadecimator in Round 4.

SHELL SHOCK

Shell Shock was given battle fatigue and eliminated from the *Robot Wars* 3 (UK) tournament in the opening round of its heat by Sir Chromalot.

SHISH-KA-BOT 1.1

Richard Wood, John Hoffman, and Clan TYFP of San Marcos, California, built Shish-ka-bot 1.1. Built on wheels and battery powered, Shish-ka-bot's primary weapon is a pneumatic spike that operates on compressed carbon dioxide. It also has two steel wood-splitting wedges mounted onto appendages that resemble arms as secondary weapons.

Shish-ka-bot was skewered in the opening round of *BattleBots*' Season 3 lightweight tournament by Trilobot.

SHOCKWAVE

A 29–16 loser to SnowFlake in the opening round of *BattleBots*' Season 3 lightweight tournament, Shockwave was built by James Arluck and Team SMC of Goleta, California.

SHORT CIRCUIT

Lorena Blalock of Fairfax Station, Virginia, built Short Circuit, which lost a crushing 41–4 decision to Little Drummer Boy in the opening round of *BattleBots*' Season 3 middleweight tournament.

SHORT ORDER CHEFF

Mark Curley and Team Bus Boys of Tecumseh, Michigan, built Short Order Cheff, which knocked out Zulu in the opening round of *BattleBots*' middleweight tournament. In Round 2, it won a 34–11 decision over Musashi, then lost a 27–18 decision to Exodus 2001 in Round 3.

SHOVIT

As in "take this job and . . ." Shovit was built by Joseph Murphy of St. Albans, Vermont. It won a 23–22 decision over T-6 before being KO'd by Garm in Round 2 of *BattleBots*' middleweight tournament.

SHRAPNEL

Shrapnel proved to be dairy-intolerant and was eliminated from the *Robot Wars* 3 (UK) tournament in the opening round of its heat by Big Cheese.

SHREDBOT

Shredded by Violator in the first round of *BattleBots*' middleweight tournament, Shredbot was built by Larry Beckman of Lake Park, Illinois.

SHRIKE

A fearsome competitor, Shrike was built by Steven Levin and Team Masada of Mountain View, California. At *BattleBots*' Season 1 lightweight tournament, Shrike was eliminated in an early round by Das Bot.

None of Shrike's three matches in *BattleBots*' Season 3 lightweight tournament lasted the full three minutes, such was the carnage. After knocking out Spin Reaper and Whirligig, Shrike was itself blasted into oblivion by Ziggo.

SHUNT

House bot on *Robot Wars: Extreme Warriors,* Shunt weighs 240 pounds and measures fifty-two inches by forty-four inches by twenty-eight inches. With exceptional pushing power, Shunt gets its juice from a prototype electric motor. It attacks other bots with both a diamond-edged ax and a scoop. Flip Shunt over and it has the ability to self-right.

SILVER BOX

Silver Box was built by Mike Regan and Ronn Holmes of Redondo Beach, California. It weighs 174 pounds, and measures forty inches by twenty-eight inches by fifteen inches. It is powered by a pair of electric motors and its main weapon is its lethal rotating disc. Though heavily armored, Silver Box has proven unreliable in combat.

In the "Robot Rebellion" episode of *Robot Wars: Extreme Warriors,* in which entrants competed for the opportunity to square off against the *Robot Wars*' house bots, Silver Box defeated Close Enough in the eliminator. At the end of that match, Silver Box pushed Close Enough onto the arena's built-in flipper and sent it flying. That victory made it one of the four entrants to combat house bots Shunt, Sir Kill-a-lot, and Matilda—the house bots losing in a tremendous upset.

SILVER BULLET

David Stevens and Team Kiss This of Yuma, Arizona, built Silver Bullet, which was designed to kill werebots (you know, bots that turn into wolves in the full moon).

At *BattleBots*' Season 3 lightweight tournament, Silver Bullet KO'd Æon 00 in the opening round before being KO'd by Patriot in Round 2.

SIR CHROMALOT

Sir Chromalot was a winner in the opening round of its heat at the *Robot Wars* 3 (UK) tournament, defeating Shell Shock before losing to Big Brother in the second round.

SIR KILL-A-LOT

A super-superheavyweight at 616 pounds, Sir Kill-a-lot is one of the *Robot Wars: Extreme Warriors* house bots that roam the arena smashing, bullying, and abusing the losers when they are down and out.

Built on track wheels, this bot measures forty-eight inches by forty-eight inches by fifty-two inches. With a head on top that wears a Darth Vader–like helmet/mask, it runs on gasoline (or petrol, as they call it in England).

Sir Killabot has two weapons, a pair of crushing claws and a lance. The arm with the claw on it is so strong that it can clamp down on smaller bots (they are all smaller bots) and throw them across the arena.

SISYPHUS

Daniel Fernandez and Team Bust-a-Bot of Houston, Texas, devised this formidable opponent, which made it to the fourth round of *BattleBots*' Season 3 lightweight tournament. It defeated Parallax 30–15, KO'd Bad Bone, and outpointed Pursuer 27–18 before being KO'd by Death by Monkeys.

THE SIX MILLION DOLLAR MOUSE

Built by Chris, Greg, and Richard Hannold of North Carolina, this expensive rodent weighs 220 pounds and measures thirty-six inches by twenty-six inches by twenty-three inches. It runs on 24-volt electric motors. The Six Million Dollar Mouse is heavily armored and its primary weapon is a spinning rotor blade.

On the "Civil War" episode during the first season of *Robot Wars: Extreme Warriors,* The Six Million Dollar Mouse represented the South. On that show, it defeated Falcon, knocked out Sobek, and then was knocked out in the finals by Manta.

In the "American Revolution: War of Independence" episode of *Robot Wars: Extreme Warriors,* in which bots from the United States were pitted against bots from the United Kingdom, the Mouse, representing the United States, was KO'd in the opening round by Great Britain's Pussycat, the eventual winner of the tournament. Pussycat's buzz saw gave the Mouse a nose-ectomy on its way to victory.

SKEWER RAT

Skewer Rat is a wedge protected by hardened aluminum armor. It gets its power from four wheelchair motors and its main weapon consists of a razor-sharp hardened tool steel skewer mounted on a 300-pound-capacity lifting arm.

That skewer is designed to pierce the armor of other robots then lift and shred. Ouch. The amazing thing about this bot is that it was entirely hand-built in under seven weeks with a budget of less than $700. Such things are still possible, kids.

The Spartan builders are automotive technology instructor Mike Fobes, age forty-one, and computer science instructor Kinsey Fobes, age forty-seven, of Federal Way and Renton, Washington. Also on the team is Kris Fobes. Skewer Rat made its competitive debut at *Robotica II.*

SKID MARK

Is it me, or are some unnecessarily gross images forced into your brain when a bot named Skid Mark fights one called Half Ass and Random?

Anyway, Skid Mark KO'd Half Ass and Random in the opening round of *BattleBots*' Season 3 lightweight tournament. It followed that victory up with another KO win over Fang. It outpointed Crashius Clay 33–12 then was KO'd by Backlash.

S.L.A.M.

Built by Lowell Nelson—along with Charles and Cheryl Booth, all of Quincy, California—S.L.A.M. is a superheavyweight bot that runs on wheels and is powered by liquid fuel.

Three years in the making, S.L.A.M.'s primary weapon is a 147-pound rotating drum with teeth.

In the *BattleBots*' Season 3 tournament, S.L.A.M. won a 24–21 decision over Armageddon, then was KO'd in the second round by Rammstein.

SLAM JOB

Scott Kincaid and Team BlackRoot of Hansen, Idaho, constructed Slam Job, which won a 33–12 decision over Negative Reinforcement in the opening round of *BattleBots*' Season 3 heavyweight tournament, then KO'd Gang Green Weasel in the second round before losing a 25–20 decision to Nightmare.

SLEDGEJAMMER

William Sauro and his team of "Fat Cats" from San Francisco, California, built SledgeJammer, which was knocked out by Bad Dog in the opening round of *BattleBots*' Season 3 heavyweight tournament.

SMD

Barry Mullins of the United State Air Force Academy in Colorado created SMD, which knocked out Caliban in the opening round of *BattleBots*' middleweight tournament before losing by KO in Round 2 to Horrifica.

SMIDSY

Smidsy was eliminated in the opening round of its heat by Rattus Rattus at the *Robot Wars* 3 (UK) tournament.

SNAKE

Snake is a superheavyweight with wheels and battery power. A veteran of *Robotica,* it has exceptional mobility for its weight class. Snake's primary weapon, also battery powered, is a pair of large steel claws.

Snake was built over four weeks by Mark Setrakian, Paul Rivera, and Peter Abrahamson of Los Angeles, California. At *BattleBots*' Season 2 superheavyweight competition, Snake made it to the second round before being eliminated by War Machine.

SNAP DRAGON

If you must name your bot after a flower, then Snap Dragon is the way to go. Built by Gregory Gancarz of San Diego, California, Snap Dragon KO'd Headtrimmer in the opening round of *BattleBots*' Season 3 lightweight tournament, then lost a 34–11 decision to Whirligig in Round 2.

SNIPE

Having been left holding the bag on a nine-day snipe hunt once, this bot is a personal favorite. Charles Haberman and Invisible Works of Bloomington, Minnesota, built Snipe, which was KO'd by I-Will-Smack-U in the opening round of *BattleBots*' Season 3 lightweight tournament.

SNOWFLAKE

Michael Mauldin and Team Toad of Irwin, Pennsylvania, built Snow-Flake, which KO'd Frozen Toast in the opening round of *BattleBots*' Season 3 lightweight tournament. It outpointed Shockwave 29–16 in Round 2 and was then melted 34–11 by Gamma Raptor in the third round.

SOBEK

Sobek was built by Glen Sears, Karen Gilmer, and Aaron Turcotte from Massachusetts. The bot weighs 216 pounds and measures seventy-five inches by forty-four inches by twelve inches. It runs on a 24-volt motor and its primary weapon is a pneumatic chomping jaw. Its main weakness in combat is that its wheels are exposed.

Sobek appeared in the "Civil War" episode during the first season of *Robot Wars: Extreme Warriors.* Sobek represented the North and won a close decision over The Revolutionist in the opening round before losing to The Six Million Dollar Mouse.

SOFO (SPACE OPERATIONS FORCE)

From the final frontier, actually Madison, Wisconsin, come Dennis and Kathleen Adams, creators of SOFO, which stands for Space Operations Force. It is a low-profile middleweight bot, built over the course of three months. It is built on wheels and has battery power. SOFO's primary weapon is a battery-operated flipper.

SOLAR FLARE

A veteran of the first season of the *Robotica* TV show, Solar Flare weighs 124 pounds. It has an aluminum frame and armor. That armor is ¼ to ½ inch thick. Its four DC motors drive coaxial wheels to give it 66 pounds of pushing force.

A custom-built control system gives Solar Flare differential steering control. Two microcontrollers linked by a serial RF system and a custom joystick and button assembly, as well as two custom speed controls for the motors, provide the control system.

Solar Flare is named after its primary weapon, a propane flamethrower.

Solar Flare was built by Mariusz Zaczek, age twenty-six, of Chicago, Illinois, and Aaron Trask, age twenty-four, of Savoy, Illinois. Zaczek is a master's student in aeronautical and astronautical engineering at the University of Illinois. He is also an Emergency Medical Technician, and works in the Flight Dynamics Branch at NASA Johnson Space Center in Houston. Trask is a Ph.D. student in aeronautical and astronautical engineering at the University of Illinois. They say that their favorite film robot is the lovable Number Five from the movie *Short Circuit.*

Before *Robotica,* the builders had competed in Jerry W. Sanders Creative Design competition at the University of Illinois at Urbana-Champaign, where they were the 2001 winners of Demolition Day.

In 2000 they won the Most Impressive Arsenal Award, and in 1999 they were winners of both Demolition Day and the Most Impressive Arsenal Award.

SON OF A MONKEY WRENCH

Son of a Monkey Wrench was designed and built by twenty-seven-year-old John Adipietro and Team Rous of Cincinnati, Ohio. John is a full-time student at the University of Cincinnati, majoring in computer engineering.

He is also a systems engineer for IBM. His hobbies include autonomous mobile robots with artificial intelligence, robot vision, combat robots, radio-controlled cars, and making things "mobile."

He likes attaching motors to scooters, bikes, Rollerblades, etc. He's also into creating Web sites. John is a member of the Seattle Robotics Society. Son of a Monkey Wrench was invited to Season 1 of the *Robotica* TV show but, unfortunately, did not work upon arrival.

John and his bot promised to be back, however—and indeed, he did get an opportunity to participate in *Robotica*'s second season. This bot's primary weapon: a "massive flamethrower!" It also boasts 500 amps of power!

John's teammates are Jeremy Daniel, age twenty-five, Wayne Eberle, and Jeff Bragg, all of Cincinnati, Ohio.

SON OF BODY PARTS

Out of the gothest of all goth cemeteries comes this bot, looking like something out of a Universal horror movie. It's ali-i-i-ive! Son of Body Parts, a battery-powered lightweight, is a veteran of *BattleBots.* It is a wedge that also boasts metal cutting blades. It was built over a three-week period by Cuong Chau of Camp Hill, Pennsylvania.

SON OF WHYACHI

A heavyweight champion, this future Hall-of-Fame bot was built by Terry Ewert of Dorcester, Wisconsin. Ewert grew up on his family's Wisconsin dairy farm. Today he is married to Elizabeth, another member of Team Whyachi, and they have four sons.

At *BattleBots*' Season 3 heavyweight tournament, Son of Whyachi defeated Shaka 29–16 in the opening round, and then scored five consecutive KO victories. In order, Son of Whyachi sent Crab Meat, Kill-o-amp, Nightmare, Mechavore, and Hexadecimator into la-la land. That took Son of Whyachi into the finals, where it defeated BioHazard by a score of 29–16 to win the Giant Nut.

Son of Whyachi uses powerful Briggs & Stratton Etek motors. The bot is the first bot to walk on feet rather than roll on wheels to win the top prize at a *BattleBots* competition. The bot is expected to move up to the superheavyweight division for future contests.

SONIC

Not to be confused with the hedgehog, this entity is 100 percent bot. At the *Robot Wars* 3 (UK) tournament, Sonic defeated Hammertron in the opening round of its heat before being eliminated by the eventual tournament winner, Chaos II.

SPARE PARTS

Bill Morse and Mark Thompson—along with Team Anthrax (hey, this was before)—of Rocklin, California, constructed Spare Parts out of, well, you know. It handily defeated Count Botula by a 36–9 score in the opening round of *BattleBots*' middleweight tournament, then was outpointed 24–21 by Deb Bot in Round 2.

SPARKY

Lee Kohn of Arlington, Virginia, built Sparky, which outpointed Bambino 39–6 in the opening round of *BattleBots'* Season 3 lightweight tournament, then lost a 34–11 decision to Wacker.

SPAWN AGAIN

From Great Britain, Spawn Again weighs 174 pounds and measures forty-nine inches by thirty-three inches by twenty inches. It runs on 4.5-kilowatt starter motors and its weapon is a pneumatic flipper. Though it is heavily armored and therefore difficult to damage, it does not have the ability to self-right itself in case it gets tipped over.

Spawn Again was built by Darren Ball, Graham Warner, and Luke Jackman.

In the "American Revolution: War of Independence" episode of *Robot Wars: Extreme Warriors,* in which bots from the United States were pitted against bots from the United Kingdom, Spawn Again, representing England, was knocked out in the opening round by Joker. Spawn Again gave a disappointing showing here, breaking down pretty much after the bout's initial contact.

SPAZ

Mike Regan and Carol and Lott Lowery of Redondo Beach, California, built Spaz, a middleweight veteran of *BattleBots.* Built on wheels and battery powered, Spaz's primary weapon is a "hammer spike." The first version of the bot took four months to build, but the current version is the result of a two-and-a-half-week revision that recently took place.

SPEED BUMP

Built by Joseph Murawski and BotBashers of Beaverton, Oregon, Speed Bump lost a 36–9 decision to Mantis in the first round of *BattleBots'* Season 3 lightweight tournament.

SPIKE

At the *Robot Wars* 3 (UK) tournament, Spike lost in the opening round of its heat to Blade. Spike also appeared in the "Civil War" epsiode during the first season of *Robot Wars: Extreme Warriors,* representing the North. Spike was KO'd in the opening round by Manta.

SPIN CYCLE

Dean Roberts and Team Squeeky Kleen of Southfield, Michigan, built this bot, which was shellacked, almost shut out, 44–1 by Can Opener in the first round of *BattleBots'* middleweight tournament.

SPINNE

Built by Nathan Barnes of Golden, Colorado, Spinne lost 38–7 in the opening round of *BattleBots'* Season 3 heavyweight tournament to Kritical Mass II.

SPINNING MAYHEM

Lowell Carpenter of Santa Rosa, California, built this bot, which lost to Shark Byte in the opening round of *BattleBots'* Season 3 heavyweight tournament.

SPIN ORBITING FORCE

Spin Orbiting Force has a long arm in front with a spinning blade at the end of it—so, in the vernacular of pugilism, it usually has a reach advantage over its opponent, a comfort zone in which it can strike but cannot be struck.

In *BattleBots'* Season 1 middleweight tournament, Spin Orbiting Force defeated the lethal Blade Runner before losing to the eventual winner, Hazard.

SPIN REAPER

Built by Ryan Hahn of Omaha, Nebraska, Spin Reaper won a 25–20 decision over Critical Condition in the first round of *BattleBots'* Season 3 lightweight tournament, then was KO'd in Round 2 by Shrike.

SPINSTER

Spinster was knocked out in the first round of *BattleBots'* middleweight tournament by Deadblow. It was constructed over a four-month period by

Grayson and Anthony DuRaine and Team Deranged of Stockton, California. This is a stompbot, built on legs. It is battery powered with a spinning weapon.

SPRING BREAKER

Spring Breaker may be named after the loud and rowdy but is designed to resemble the silent and surreptitious. Its design is based on that of the futuristic F-117 Stealth fighter.

The bot has an aluminum TIG-welded frame. Its skin is ⅛ inch aluminum. Its shape tends to diffuse any blows it receives. It is extremely hard to get a clean shot at Spring Breaker. Painted on its body is a bald eagle and scenes of Daytona Beach, Florida, where it was born.

The bot's father is twenty-nine-year-old Rich Riendeau. He says that it took eight weeks to build, and it is the first bot he has ever constructed.

According to Riendeau: "Prior to this I have done full-scale electric and hybrid vehicle conversions. A robot to me is nothing more than a scaled-down electric car with a funny shape."

Riendeau says that his favorite showbiz bot is K-9 from the British *Dr. Who* television series.

SQUIRRELLY D

The nuts this bot gathers need bolts. Squirrelly D is the product of main builder Patrick King and BC Racing of Atlanta, Georgia. In combat, this rodential bot was ousted 35–10 in the opening round of *BattleBots*' middleweight tournament.

STEALTH

At the *Robot Wars* 3 (UK) competition, Stealth defeated Death Warmed Up in the opening round of its heat, but was then eliminated in the second round by the eventual heat winner, Hypnodisc.

STEALTHBOT—THE TALLAHASSEE LASSEE

Great name, lousy record. Stealthbot—The Tallahassee Lassee was eliminated in the opening round of *BattleBots*' middleweight tournament when it was KO'd by another rhyming-name bot, Orlando Commando.

STEEL AVENGER

At the *Robot Wars* 3 (UK) tournament, Steel Avenger defeated Schumey II in the opening round of its heat, but lost in the second round to Diotoir.

STEEL REIGN

Steel Reign, built by Brady Davis of Arlington, Texas, lost in the opening round of *BattleBots*' Season 3 superheavyweight tournament, coming up on the short end of a 23–22 squeaker to Hoobot.

STEG-A-SAW-US

At the *Robot Wars* 3 (UK) competition, Steg-a-Saw-Us won its heat with victories over Drac's Revenge, Henry, and Napalm. In the semifinal heat, Steg-a-Saw-Us made it into the tournament's final four, defeating Gravedigger and Beast of Bodmin before finally being eliminated by Hypnodisc.

STEG 2

At the *Robot Wars* 4 (UK) tourney, Steg 2 won its heat with victories over Iron Awe and Mortis.

STEWBOT

John Hargrave was the main builder of Stewbot, a lightweight veteran of *BattleBots,* but he had help from Moses Blumenstiel and Al Natanagara, all of Cambridge, Massachusetts. The first thing you notice about Stewbot is that its body is constructed entirely of LEGO. It is a stompbot, built on legs. It took Hargrave and his team six weeks to build.

STING II

Sting II was defeated by Diotoir in the opening round of its heat at the *Robot Wars* 3 (UK) tournament.

STINGER

From England, Stinger weighs 182 pounds and measures twenty-eight inches by twenty-four inches by twenty-five inches. It runs on two 24-volt electric motors and its weapon is a bludgeoning spike. It was built by the team of Kevin Scott, Tim Mann, and Karl Skinner.

In the *Robot Wars* 3 (UK) tournament, Stinger lost its opening match to Mace II.

In the "American Revolution: War of Independence" episode of *Robot Wars: Extreme Warriors,* in which bots from the United States were pitted against those from the United Kingdom, Stinger, representing the Queen, lost in the opening round to the ravenous jaws of that hungry yank, General Chompsalot.

STINGRAY

It's a cool fish, a cool bike, a cool car, a cool British TV show featuring supermarionation (really cool marionettes)—and a so-so bot. Stingray the bot lost a 37–8 decision to Zion in the opening round of *BattleBots'* middleweight tournament. It was built by Buck McGibbony of Los Angeles, California.

SUBJECT TO CHANGE

In *BattleBots'* Season 1 middleweight tournament, Subject to Change, a low wedge, looked like a sure winner in the opening round. It scored heavily in the early going in its bout against Pressure Drop, using its lower center of gravity to its advantage and driving Pressure Drop into the spikes that line the walls of the arena. Just when it looked as if Subject to Change was going to cruise to victory it suffered an unexpected mechanical failure.

SUBJECT TO CHANGE WITHOUT REASON

Built by Randy Eubanks, David Ray, and Team Boomer of Long Beach, California, Subject to Change Without Reason lost via KO to Complete Control in the opening round of *BattleBots'* Season 3 tournament.

Subject to Change Without Reason is built on wheels and battery powered, with a battery-powered pickax as its primary weapon. It took Eubanks and his team one month to build. And, of course, it could be completely different the next time you see it—just for the heck of it.

SUBLIME

Jerome Miles and Team Duct Tape of Provo, Utah, built Sublime, which outpointed Whirl Wep X25 by a score of 23–22 in the opening round of

BattleBots' Season 3 lightweight tournament, then was KO'd in Round 2 by Sallad.

SUBLIME II

Son of Sublime. Like its dad, Sublime II was built by Jerome Miles, Julie Babb, Jon Babb, and Team Duct Tape of Provo, Utah. It was built over a four-week period. It moves on wheels and is battery powered. A battery powered actuated grabbing pincer is its primary weapon. It is small, even for a lightweight, and therefore difficult for its opponents to hit.

SUICIDAL TENDENCIES

With tracked wheels like those of a tank, Suicidal Tendencies is battery powered. Its weapons, a pickax and wedge scoop, are powered by compressed gas. This bot has the ability to run inverted. Its wedge scoop can lift 250 pounds. The ax, used in conjunction with the lifter, allows Suicidal Tendencies to crush its opponents.

The bot has the ability to climb stairs and the power to pull a Land Rover. It was built over the course of seven months by Charles Binns, Andrew Jeffery, and Richard Hutchinson of Derby, England.

At *BattleBots*' Season 2 heavyweight competition, Suicidal Tendencies was flipped onto the kill saws and KO'd in the opening round by the eventual winner, BioHazard.

In *Robot Wars* 3 (UK) tournament, it defeated Forklift's Revenge in the opening round of its heat, then outpointed Raizer Blade in the second round. In the heat finals, Suicidal Tendencies was eliminated from the competition by Mace II.

SUMP THING

Sump Thing was eliminated from the *Robot Wars* 3 (UK) tournament in the opening round of its heat by the eventual heat winner, Pit Bull.

SUNSHINE LOLLIBOT

This bot's name is derived from the classic Leslie Gore song, "Sunshine, Lollipops and Rainbows," which appears on the soundtrack of the major motion picture *Ski Party.* It's also a damn fine bot.

At *BattleBots*' Season 3 middleweight tournament, Sunshine Lollibot scored four victories before it was eliminated. It KO'd Nemisis in the opening round, won a 30–15 decision over Metallic Mayhem in Round 2, KO'd Ravager in the third round, KO'd Alabama Slammer in the fourth round, and then lost by a KO to T-Minus.

Sunshine Lollibot was built by Andrew Miller and Team Sunshine of Quilcene, Washington.

SUPER CHIABOT

It grows! Super Chiabot is built on wheels and battery powered. It multiplies! It is the mother ship for a set of three smaller bots known as Chia's little helpers. Its primary weapon is a battery-powered spinning disc.

This middleweight veteran of *BattleBots* was built by Will Wright, Justin McCormick, and Gabriel Heilig-Yochim of Orinda, California.

SUPER ORBITING FORCE (SOFO)

Super Orbiting Force (sometimes known as Sofo) weighs 113 pounds, runs on wheels, and has a drive and weapon that are both battery powered. A saw-blade arm is its primary weapon.

Builder Michael Winter of Madison, Wisconsin, is the leader of the Robot Action League, a team that sounds like it should have its own comic book, with Dennis Adams being the other "Action" guy.

SUPER SHARK

Built by Dave Sandel of Shirley, New York—who calls his shop Blood, Sweat and Spears—Super Shark lost in the opening round of *BattleBots*' Season 3 superheavyweight tournament to Junkyard Cog.

SURGEON GENERAL

Here's a bot that is definitely hazardous to the health of other bots. It is wheeled and battery powered. Its primary weapon is high-speed rotary blades. It is reversible so that it functions equally well when right-side up or when upside down.

It was built over a five-week period by Korey Kline and Julio Roqueta of Miami, Florida, with financial aid from Florida International University.

In *BattleBots*' Season 3 heavyweight tournament, Surgeon General KO'd Mjollnir in the opening round, KO'd KillerB in the second round, and then lost, also by KO, to Mechavore in Round 3.

THE SWARM

The Swarm was a joint effort, built by Stephen Felk, Jim Smentowski, and Paul Mathus of San Francisco, California. It lost in the opening round of *BattleBots*' Season 3 superheavyweight tournament to Dawn of Destruction, outpointed by a measly 23–22 margin.

SWORD PLAYER

In *Robotica*'s Season 1 tournament, Sword Player made it past the preliminary rounds but was eliminated by Ram Force in the Fight to the Finish.

T

T II

Not *Tea for Two,* just T II. At the *Robot Wars* 3 (UK) competition, T II defeated Vector in the opening round of its heat before succumbing in the second round to Darke Destroyer.

T-6

Ryan Bialaszewski of Santa Clara, California, built T-6, a middleweight that lost its first match by only one point (23–22) to Shovit in *BattleBots*' Season 3 tournament.

TAKATAKATAK

TakaTakaTak weighs 210 pounds and made its competitive debut at *Robotica II.* It has four-wheel differential steering and its two front wheels have automotive-style steering, so it has great agility for its weight.

Its weapons are both CO_2-powered pneumatic weapons, 600-pound lifting forks and a pair of bashers. Its chassis is constructed of ⅜-inch T6 aluminum. It is powered by two 1.6-horsepower wheelchair motors.

It may be agile but it doesn't exactly dash. Top speed: five miles per hour. TakaTakaTak was built by engineer and project manager Don Gibson, age forty-five, along with Team Skid Mark: Jenny Gibson, and Doug Gibson, age thirteen. They are from Abbotsford, British Columbia, Canada.

Don and Doug say that their favorite Hollywood bots are the microbots from the movie *Batteries Not Included.* Their hobbies are microelectronics, machine shop, and multiplayer Internet FRP games.

TANTRUM

Paul Ventimiglia and his Demolition Team of Wayne, New Jersey, built Tantrum, which lost a 27–18 decision to Manic Aggressive in the opening round of *BattleBots*' Season 3 lightweight tournament.

TARKUS

Tarkus was constructed by David Cain of Tujunga, California. The bot lost a 27–18 decision to Parasitic in the opening round of *BattleBots*' Season 3 heavyweight tournament.

TAZBOT

Tazbot is a heavyweight bot, weighing 207 pounds. Battery powered, with wheels, Tazbot uses a customized pickax as its primary weapon. Tazbot has a 360-degree rotating turret and an interchangeable tail.

The main builder is Donald Hutson of San Diego, California. The other members of Team Tazbot are David Cook, David Blakey, Dawn Bernstein, Paul Remond, and Mike Moore.

Tazbot won a 34–11 decision over BattleRat in the opening round of *BattleBots*' Season 3 heavyweight tournament. It followed that up with a 38–7 decision over GoldDigger before losing to BioHazard in Round 3 by a score of 27–18.

TECHNO DESTRUCTO

One of the toughest of the big boys, Techno Destructo was built by Sean Irvin of Melbourne, Florida. In *BattleBots*' Season 3 superheavyweight tournament, Techno Destructo won four decisions before being knocked out in the fifth round. It defeated the Little Blue Engine That Could by a score of 23–22, Metalhead 32–13, Cuad the Crusher 23–22, and Atomic Wedgie 35–10 before losing via snoozing to Vladiator.

TECHNOFOOL

The creation of Andrew Robinson of Sacramento, California, Technofool was sent packing in the opening round of *BattleBots*' Season 3 heavyweight tournament by Mechavore.

Built on wheels out of tractor-trailer parts and battery powered, Technofool's primary weapon is a spinning drum with twin balls and chains. The weapon is powered by compressed gas.

The bot's sturdy platform is comprised of a twenty-four-inch aluminum truck wheel and hub mounted on the steering spindle. According to Robinson, Technofool was under construction for four months.

TECHNOPHOBIC

Technophobic battled irrational fear and emerged victorious in its opening-round match in the *Robot Wars* 3 (UK) tournament, defeating Abaddon, then lost in Round 2 to the eventual heat winner, Pit Bull.

TENTOUMUSHI

Resembling a lady bug, this lightweight (54 pounds) was built by Lisa Winter and Team Action League. The bot was the winner of the Funniest Robot award at *BattleBots* Season 1.

Tentoumushi tossed Sallad right out of the competition at *BattleBots*' Season 1 tournament. The bug was subsequently squashed by Alpha Raptor.

TERMINAL FEROCITY

At the *Robot Wars* 3 (UK) tournament, Terminal Ferocity was defeated in the opening round of its heat by Facet.

TERRAPIN

With the protection of its tortoise shell, Terrapin defeated Vercingetorix in the opening round of its heat at the *Robot Wars* 3 (UK) competition, then was defeated and eliminated in the second round by Invertibrat.

TETANUS

Tipping the scales at 220 pounds, Tetanus has four-wheel drive and is constructed of steel, Lexan, and ballistic nylon. Tetanus uses two 2.5-horsepower motors that came off an electric riding mower.

As a weapon, it has three removable front spikes. They can be interchanged with a set of conical steel blades spinning at 3,800 revolutions per minute, which are driven by a 5-horsepower electric motor.

The team, known as Browns Mountain Lab, consists of electronics technician Kevin Evans, age thirty-eight, and nonprofit consultant Erin Evans, age thirty-seven. They come from Hawkins County, Tennessee.

Tetanus made its competitive debut at *Robotica II.*

TEXAS STAMPEDE

Built by Thomas Robertson of Austin, Texas, Texas Stampede lost a 27–18 decision to Blood Moon in the first round of *BattleBots*' Season 3 lightweight tournament.

THAT GOOD R&D

Jason Stokes of Tewksbury, Massachusetts, created this bot, which was KO'd by Hexadecimator in the opening round of *BattleBots*' Season 3 heavyweight tournament.

THERMODOR

At the *Robot Wars* 3 (UK) competition, Thermodor laid some high pressure on Plunderstorm in the opening round of its heat, but was then eliminated in the second round by Scutter's Revenge.

THING II

At the *Robot Wars* 3 (UK) competition, Thing II was the winner of its heat, scoring consecutive victories over Prizephita, All Torque, and Cerberus. In the semifinal heat, however, Thing II bowed out in the opening round, losing to Panic Attack.

THOR

Thor's main weapon is a carbon-dioxide-powered flipping/lifting arm. It was built by David Roy, age fifteen, and Robert Simms, age forty-two, of Los Angeles, California. David is a student and Robert is a welder.

Their hobbies are high-power rocketry, Tae Kwon Do, robotics, computers, go-pedding, building things, welding, and radio-controlled cars. Thor made its competitive debut at *Robotica II.*

THWAK

Thwak was built by Mike Regan and Team Vicious of Redondo Beach, California. In *BattleBots'* Season 3 middleweight tournament, Thwak defeated Psychotron 23–22 in the opening round, whacked Mobster in Round 2, outpointed Traxx 36–9 in the third round, then lost by KO to Twin Paradox.

TIGER CAT

Built by the Dave Bandels, Sr. and Jr., of Shirley, New York, Tiger Cat weighs 220 pounds and measures forty inches by twenty-six inches by twenty inches. It runs on two electric motors and has a pneumatic ax as its primary weapon.

The feline's weakness (especially when it is competing in the *Robot Wars* arena) is that it is flammable. In the "Civil War" episode during the first season of *Robot Wars: Extreme Warriors,* Tiger Cat represented the North. Taking on Manta in the opening round, Tiger Cat appeared to lose a piece every time there was contact and, naturally, it burned.

TIMMY

Mike Hertzig and Team Truly Unruly of Orangeville, California, put together Timmy, which scored a pair of KOs over Mr. Sucko and Kinetic Kill Vehicle before it was eliminated from *BattleBots'* Season 3 middleweight tournament.

T-MINUS

An awesome bot, 115-pound T-Minus was built by Zander Rose and Reason Bradley, and Inertia Labs of Sausalito, California. Its primary weapon is a super-powered pneumatic flipper.

T-Minus scored five consecutive KO victories in *BattleBots'* Season 3 middleweight tournament before being KO'd itself. Its victims in that tourney were, in order, Halo, Kraken, Cheap Shot, BumperBawt, and Sunshine Lollibot. It was defeated by T-Wrex, but nonetheless was awarded the Coolest Robot designation for the tournament.

TOBOR RABIES

Hey, backward that spells SEIBAR ROBOT! Huh? Anyway, Tobor Rabies is a middleweight veteran of *BattleBots,* built on wheels and battery powered. Its weapons include spikes, wings, saws, and its own ramming power. Toby was built by Darrin, Chance, and Spenser McLand of The Colony, Texas.

TOE CRUSHER

A Christian Carlberg bot, the 55-pound Toe Crusher advanced as far as the semifinals in the *BattleBots*' Season 2 lightweight competition, losing at the two-minute, fifteen-second mark to Backlash. If you ask Carlberg how the bot got its name, he shows you his toe.

"I was standing in front of the bot when I pressed the wrong button and its hammering spike did the rest," Carlberg says of his battery-powered, built-on-wheels creation.

Carlberg built Toe Crusher over a one-month period using a lifetime of knowledge. He had the assistance of Jessica Lehrbaum, who, like Carlberg, is from Santa Monica, California.

In the Season 3 tourney, Toe Crusher lost a 23–22 decision in the opening round to Dr. Inferno Jr. Toe Crusher also competed in the 1999 BotBash.

TOE CUTTER

At the *Robot Wars* 3 (UK) tournament, Toe Cutter defeated Purple Predator in the opening round of its heat, then lost in the second round to the eventual heat winner, Panic Attack.

TOOL CHOP

Tool Chop was built by Luis Floriano and Team Wizard of El Paso, Texas. In *BattleBots*' Season 3 heavyweight tournament, this bot lost in the opening round by a 28–17 score to Center Punch.

TORNADO

The version of Tornado that competed in *Robot Wars* 4 weighed 78.2 kilograms. It measured 85 by 70 by 25 centimeters. Its batteries were 36-volt, 14ah and it ran at a top speed of ten miles per hour.

For Series 5 of *Robot Wars,* Tornado weighed 100 kilograms. It measured 90 by 75 by 24 centimeters. Its batteries were 36-volt, 16ah and the top speed remained at ten miles per hour.

Its main weapon is a 7-kilogram disc that revolves at 2,500 revolutions per minute. Tornado's chassis is constructed almost entirely with 1-millimeter-wall 1-inch box section steel, welded together with a MIG welder. Some key parts are made of thicker wall (up to 3 millimeter) material. Two-millimeter-thick plates are used for the front part of the motor mounts. M6 bolts are used everywhere possible.

For Series 4, the armor was made of 6-millimeter polycarbonate (tuffak). It was fastened to the chassis with M6 countersunk bolts, which screwed into M6 plain nuts welded to the chassis. The access flap was held down using neodymium magnets.

For Series 5, the bot was laminated with titanium over critical areas. The link access door was moved to the back of the robot to keep it away from the weapon and to make it accessible regardless of which side of the bot was up.

Tornado's first-stage gearing used steel MOD2 gears from HPC. A twelve-tooth pinion was fitted to the motor with a solid 5-millimeter dowel pin, and the forty-eight-tooth spur gear was bolted to a machined-down go-kart clutch. This gave a first-stage gearing ratio of 1:4.

For Series 4 the intermediate gear assembly ran on the roller bearing contained in the go-kart clutch. This sat on a custom spacer on a 17-millimeter silver steel axle. Since this has a tendency to move during fights, and the spacer wore very quickly, for recent tournaments, Tornado changed to a longer axle with a pair of bearings.

The radio gear is a 40Mhz Hitec Ranger III transmitter coupled to a Hitec eight-channel dual conversion receiver. The transmitter and receiver can be modified to transmit and receive a full seven channels. The standard seven-channel chips are used on the circuit boards inside the units. The extra channels are disabled. As long as the aerial was adequate, no interference problems were found with this setup. The aerial was made removable so it could be replaced if damaged, and that was a good thing, as the aerial was damaged repeatedly. For Series 5, the aerial was a piece of piano wire mounted at the rear of the robot.

At the *Robot Wars* 4 (UK) tourney, Tornado—though unseeded—defeated Kater Killer in the opening round of its heat, took out Beserk II in the second round, and defeated Gemini in the heat finals.

TORO

A wheeled and battery-powered bot, Toro was constructed in the Inertia Labs of Sausalito, California, by Reason Bradley, Alexander Rose, David Estrich, Michael Scully, and Bobby Besunner of Sausalito, California.

Toro advanced to the second round before being eliminated by Atomic Wedgie in *BattleBots*' Season 2 superheavyweight competition.

The wheeled and battery-powered Toro KO'd World Peace in the opening round and Phere in the second round of *BattleBots*' Season 3 superheavyweight tournament before losing a 27–18 decision to Minion in the third round.

TORTOISE

Tortoise was built in 700 hours by Ian Burt and Paul Boucher of Edina, Minnesota. Built on wheels and battery powered, Tortoise is a superheavyweight veteran of *BattleBots.* Its main weapon is its dual-purpose kinetic energy shell, which is made out of rotating steel. The shell not only functions to protect the bot from an opponent's attacks, but it serves as an offensive weapon as well.

Tortoise spins on a fixed shaft like the wheel of a car. The motor has been modified from a car starter. Its special features include fully digital and solid-state custom electronics. These yield an amazing dual 300-plus amp capacity for fully proportional control of two electric motors. The overall power controlled through these dual H-bridges can operate two 5-horsepower electrical motors.

Tortoise was KO'd in the opening round of *BattleBots*' Season 3 superheavyweight tournament by Minion.

TOTAL CHAOS

This bot, which was KO'd in the opening round of *BattleBots*' Season 3 heavyweight tournament by Nasty Overbite, was built by Mike Hinson of St. George, Utah.

TOWERING INFERNO

Built by Jason Bardis and the Inferno Lab of Santa Barbara, California, Towering Inferno performed well in *BattleBots*' Season 3 heavyweight tournament. In the opening round, it beat Nerd Killer by a 23–22 score. It KO'd Bad Dog in Round 2 before losing via KO to Frenzy in the third round.

TOWER OF POWER

Tower of Power lost a thisclose 23–22 decision to Gungnire in Round 1 of *BattleBots*' Season 3 middleweight tournament. Its construction was engineered by Dan Haeg of Pittsburgh, Pennsylvania.

TRACIE

Tracie was the winner of its heat at the first *Robot Wars* UK series, but lost in the finals to Roadblock.

TRAINWRECK

Everyone loves a Trainwreck, right? Unfortunately, this Trainwreck lost a 35–10 decision to Bouncing Betty—who was a bit of a train wreck herself—in the first round of *BattleBots*' Season 3 middleweight tournament.

TRAVESTY

Doug Jones of Fallbrook, California, built Travesty, which scored a KO win over Necronomibot in the opening round of *BattleBots*' Season 3 middleweight tournament, then lost a 26–19 decision to Cerberus in Round 2.

TRAXX

This team is called Restless Nights. Its leader is Daniel Goff of Willits, California, who was the main builder of Traxx, which won a pair of decisions before being eliminated from *BattleBots*' Season 3 middleweight tournament. Traxx scored a 27–18 win over Goathammer and a 24–21 decision over WindChill before losing a 36–9 decision to Thwak.

TRICERABOT

Tricerabot was built by Mike Morrow, Ron Ender, Nicole Morrow, and Team Juggernaut of Corvallis, Oregon. This neoprehistoric machine weighs 220 pounds and measures sixty-five inches by thirty-six inches by fourteen inches. It runs on electric power and its main weapon is its pair of armor-piercing horns.

In the "Robot Rebellion" episode of *Robot Wars: Extreme Warriors,* in which entrants competed for the chance to fight the seemingly indestructable *Robot Wars*' house bots, Tricerabot defeated The Green Mouse in the eliminator round and earned the right to, along with three other winners, take on, battle-royale-style, house bots Shunt, Matilda, and Sir Kill-a-lot, a melee that was unexpectedly lost by the larger house bots.

In that same show's Tag Team Terror Tournament, Tricerabot teamed up with Rosie the Riveter. They lost their opening match to Tut Tut and Drillzilla, but emerged victorious in the battle of the losers that followed, defeating Joker and Falcon.

TRIDENT

At the *Robot Wars* 3 (UK) tournament, Trident won its opening heat. In the opening round, it defeated the unpronounceable Twn Trwn. Victories over Victor II and Dreadnought followed. In the semifinal heat, Trident lost in the opening round to the eventual tournament winner, Chaos II.

TRIG TERROR

Trig Terror is a middleweight veteran of *BattleBots.* It was built by Daniel Rupert, Brad Greenspan, George Marshall, and Kevin Graham of San Diego, California. Built on wheels and battery powered, Trig Terror attacks with saws.

TRILOBOT

An exceptional bot, built by Phil Putnam and LungFish Technologies of Houston, Texas, Trilobot made it to the fifth round of the *BattleBots*' Season 3 lightweight tournament. It outpointed Botulistic 24–21, KO'd Aries and Shish-ka-bot 1.1, outpointed No Tolerance III 26–19, then was KO'd by Mouser Mecha-Catbot.

TRIMANGLE

Trimangle was built at Loki Robotics of Miami, Florida, under the supervision of Nola Garcia. At *BattleBots*' Season 3 superheavyweight tournament, Trimangle defeated Battle Scar 27–18 in the opening round before losing to Hoobot 27–18.

TRIPULTA RAPTOR

Built on wheels and battery powered, Tripulta Raptor was constructed by Chuck and Robert Pitzer, Chris Harriman, James Wilkinson, and Kelly Walterschied of Chandler, Arizona, over a two-week period. This dinosaur's main weapon is a hydraulic jaw.

Tripulta Raptor knocked out Orion in the opening round of *BattleBots*' Season 3 heavyweight tournament before losing in Round 2 via KO to KillerB.

TRITEROBOT

At the *Robot Wars* 3 (UK) competition, Triterobot was eliminated in the opening round of its heat by the eventual heat winner, Evil Weevil.

TURBO

Built over a two-month period by Eddy Ampuero, and Alex Espinosa of Miami, Florida, Turbo, a middleweight veteran of *BattleBots*, is built on wheels and battery powered. Its main weapons are its provocatively monickered "sweet knockers." Turbo's armor is made of titanium that came from a jet aircraft engine.

Turbo was KO'd in the opening round of *BattleBots*' Season 3 middleweight tournament by Double Agent.

TURTLE ROAD KILL

A spinner, Turtle Road Kill lost in *BattleBots*' Season 1 middleweight tournament to the eventual winner, Hazard.

T.U.S.K.

John Patrick and Team Saber of Crystal Lake, Illinois, built T.U.S.K., which lost a 32–13 decision to Blood Moon in the opening round of *BattleBots*' Season 3 lightweight tournament.

TUT'S REVENGE

At the *Robot Wars* 3 (UK) competition, Tut's Revenge lost in the opening round of its heat to Beserk II.

TUT TUT

Tut Tut is a pyramid-shaped bot built by Chris Harriman and Team Cyclone of Chandler, Arizona. It weighs 206 pounds and measures forty inches by forty inches by twenty-six inches. It is powered by two electric motors and its primary weapon is a pneumatic ax that has proven to be highly effective in combat. Once toppled, Tut Tut is unable to self-right.

In the "Robot Rebellion" episode of *Robot Wars: Extreme Warriors*, in which entrants competed for the opportunity to grapple with the awesome *Robot Wars*' house bots, Tut Tut pushed a bot named The Green House into the arena's built-in pit in the eliminator round. Tut Tut became one of four bots to take on house bots Shunt, Matilda, and Sir Kill-a-lot in a battle royale final, which the house bots lost in a major upset.

With Drillzilla, Tut Tut won the Tag Team Terror Championship at the *Robot Wars: Extreme Warriors* Season 1 tournament. On their way to victory, this team defeated Rosie the Riveter and Tricerabot, and then General Chompsalot and Run Away.

TWIN PARADOX

I'm not geeky enough to get this joke. The originators are Michael Neese and Skunk-Tek of Charlotte, North Carolina.

Twin Paradox, despite its vague name, is one helluva bot, winning five bouts before being eliminated in *BattleBots'* Season 3 middleweight tournament. Twin Paradox hit Bot-a-Bing, raced past Fast Forward, numbed Lobotomy, smacked Thwak, and buried Deadblow before being KO'd in the sixth round by Sabotage.

TWISTER

Charles Steinkuehler and Team Brute Force of Topeka, Kansas, built Twister, which lost a 32–13 decision to Lobotomy in the opening round of *BattleBots'* Season 3 middleweight tournament.

TWN TRWN

I'm taking up a collection to buy this poor bot a vowel. Well, Twn Trwn may be pronounceability-challenged, but it was easy to defeat in the opening round of its heat at the *Robot Wars* 3 (UK) tournament, where it was eliminated by Trident.

T-WREX

Bang a gong! T-Wrex did very well at *BattleBots'* Season 3 middleweight tournament. Built by Brent Regan and Regan Designs of Coeur d'Alene, Idaho, T-Wrex KO'd Scrap Metal in the opening round, outpointed Pokey 35–10 in Round 2, KO'd Pack Raptors in the third round, KO'd Scrap Daddy's Flipshanker, won a 24–21 decision over Bad Attitude, KO'd T-Minus, then lost to the eventual winner, Hazard, in the semifinals.

U

ULTOR

Ultor made a decent showing at the *Robot Wars* 3 (UK) tournament, winning its opening-round match over Flipper, outscoring Bumble Bot in the second round, then losing to the eventual heat winner, Big Brother, in the heat semifinals.

ULTRAVIOLENCE

UltraViolence, a combatant in *Robitica III,* was built by Team Suspect (David Gilson, Arthur Heckman, Jeff Cesnik, Billy Boden), the same folks who gave you the *Robot Wars'* bot known as Manta. The results of *Robotica III* are not available as we go to press, but judging from Team Suspect's past combat performance, UltraViolence should be ultra-tough to bear.

UNDERTAKER

Can you dig it? At the *Robot Wars* 3 (UK) competition, Undertaker was eliminated in the opening round of its heat by Panzer.

UNDERTOW

Undertow—built by Bryan Boucher of Pembroke, Michigan—was KO'd in the opening round of *BattleBots'* Season 3 lightweight tournament by Wedge-o-Matic.

UNLISCENSED CHIROPRACTOR

Built by Will Tatman and Team In-Theory of Virginia Beach, Virginia, Unlicensed Chiropractor was KO'd in the opening round of *BattleBots'* Season 3 lightweight tournament by Death by Monkeys.

UPLIFTING EXPERIENCE

Uplifting Experience was constructed by Bob Schneeveis and Team Snowhite of Palo Alto, California. In *BattleBots'* Season 3 middleweight tournament, Uplifting Experience won a pair of easy decisions before being eliminated. It outscored Dispose-All 37–8 and Crouching Puppy, Hidden Kitten 42–3, then was defeated 29–16 by Zion.

V

VECTOR

At the *Robot Wars* 3 (UK) competition, Vector lost to T II in the opening round of its heat.

VELOCIRIPPA

At the *Robot Wars* 4 (UK) tourney, Velocirippa lost to Robochicken in the opening round of its heat.

VELOCITY 1

Steve Yoden and Team Velocity of Piedmont, California, produced Velocity 1, which lost speed and was outpointed by a score of 32–13 in the opening round of *BattleBots*' Season 3 middleweight tournament by Gary Gizzmo.

VENUS PROBE

Winner of first place in the 12.5-kilogram class at BotBash 1999. Venus Probe was built in nine weeks at a cost of $1,411. It is driven by two Astroflight Cobalt 40 motors and receives its power from four six-cell nicad packs, 1,800mah each. It is controlled by a three-channel PCM radio. Its primary weapon is a pneumatic flipper/lifter, which is powered by a 12-gram carbon-dioxide cartridge. The frame is made of ½-inch welded square tube steel and 3-millimeter carbon fiber provides the armor.

VERCINGETORIX

At the *Robot Wars* 3 (UK) competition, this bot was eliminated in the opening round of its heat by Terrapin.

VERMICIOUS KENID

The brainchild of Pat Rainsberry and Precision Engineering of Berkeley, California, Vermicious Kenid outpointed Land Mine 26–19 in the opening round of *BattleBots*' Season 3 lightweight tournament, KO'd Mad Cow in Round 2, and then lost a 30–15 decision to Crashius Clay in the third round.

VICTOR II

At the *Robot Wars* 3 (UK) tournament, Victor II scuttled Scarab in the opening round of its heat, but then was eliminated in the second round by the three-pronged attack of the eventual heat winner, Trident.

VILLAGE IDIOT

Village Idiot is a strong middleweight bot that advanced as far as the fifth round during *BattleBots*' Season 3 (2001). It KO'd Bang, Deb Bot, and Chia-bot, along with winning a 30–15 decision over Eye of the Storm, before it was KO'd in the fifth round by Eraser. Village Idiot was built by Jon Autry and Team RV of Jackson, Tennessee.

VIOLATOR

The product of Eric Molitor and TuxBot of Dallas, Texas, Violator KO'd Shredbot in the opening round of *BattleBots*' Season 3 middleweight tournament, advanced past the second round because of a forfeit, then lost in Round 3 by KO to F5.

VIPER

A veteran of *Robotica,* Viper weighs 206 pounds. It has a steel frame with bulletproof Lexan armor on the top and bottom. It features six-wheel drive and is powered by two large wheelchair motors. Its primary weapon is its wedge shape.

Viper was built by a fifty-one-year-old electronics consultant named Bob Allen, who says that Robby the Robot from the movie *Forbidden Planet* is his favorite bot of all time.

In *Robotica*'s first season, Viper advanced to the Final Six, where it was matched up against JuggerBot in the preliminary rounds. Viper lost all three events and failed to advance to the finals.

VLADIATOR

King of the superheavyweight hill. The bigger, younger brother of Vlad the Impaler, Vladiator was the winner of *BattleBots*' Season 3 superheavyweight tournament. Vladiator has a symmetrical drivetrain, making it invertible.

Vladiator can really scoot, too, thanks to Lynch high-end electric motors, providing it with a top speed of thirty-five miles per hour. The bot's weapon is a combination spike/lifting arm. Accepting the big nut was head Vladmeister Gage Cauchois of Oakland, California.

Cauchois is one robot builder who has, thanks to excellent sponsorship from Loctite, been able to quit his day job so that he can dedicate all of his time to building metallic warriors.

Cauchois says that both of his bots have something in common: "They are high-speed forklifts."

On its way to the top, the rookie Vladiator defeated Juggernaut 28–17, Hammertime 30–15, Revision Z 30–15, Techno Destructo by KO, Diesector 24–21, and, in the finals, Minion 26–19.

VLAD THE IMPALER

Another bot by Gage Cauchois, Gregg Walker, and the Vladmeisters of Oakland, California. Vlad the Impaler, named after the historical figure who best resembles the fictional Count Dracula, is wheeled and battery powered.

Its primary weapon is a drag-racing forklift that functions with compressed gas. When tipped over, Vlad the Impaler has the ability to move itself back right-side up. It is heavily armored. Vlad, which took 2,000 hours to build, won the Giant Nut in the heavyweight division at the *BattleBots'* Season 1 tournament. At that same competition it was voted Most Vicious Bot.

The following season, Vlad defeated Voltronic in the semifinals. In the finals it faced BioHazard but failed to win back-to-back nuts, as BioHazard twice flipped Vlad to take the victory.

In BattleBots' Season 3 tournament, Vlad lost a 24–21 decision to Mechavore in the opening round. Vlad the Impaler competed for six years before being destroyed in its final match.

Gage Cauchois commented: "Vlad the Impaler was built for a different era of robotic combat. In a way, I was relieved to see its total destruction."

VOID

Void was built by Chris Hannold and Litewav Robotics of Salisbury, North Carolina. It is a heavyweight, tipping the scales at 210 pounds. It is driven by two 24-volt military surplus motors, powered by two 12-volt, 17ah and two 12-volt, 4ah SLAs.

Its armor consists of ⅛-inch and ¼-inch stainless steel and ½-inch Lexan. It is radio controlled by Futaba six-channel FM remote and four QD 120-amp speed controllers.

Void's primary weapon is a 9½-inch stainless steel spike that is ¾-inch in diameter. It is mounted on a ¼-inch-thick piece of flat stainless steel that was cut out with a laser.

Void also has an arm weapon. The arm is almost five feet long and has three black 8-inch cutting disks mounted on its end. Those discs are chain driven from a 1.4-horsepower motor mounted near the base. The arm is raised and lowered by dual linear actuators. Each actuator can lift 450 pounds and sustain 750 pounds. The arm can turn Void right-side up in case it gets flipped over in combat. There is a mechanism so that the arm can be mounted in different spots and attack from the rear or the sides, depending on the foe.

The base, actuators, and cutter drive motor are protected with polycarbonate.

VOLTRONIC

Voltronic was constructed by Stephen Felk of San Francisco, California. It is wheeled and battery powered. Its primary weapon is an articulated lifter bar, which is also battery powered. That bar rises out of the bot's body—the first thing that sticks out are three teeth made of sharpened steel. These are designed to get under the body of the opponent.

At *BattleBots*' Season 2 heavyweight competition, Voltronic racked up a series of victories, bulldozing Dreadbot in the opening round, defeating Bigger Brother in Round 2, and taking out FrostBite in the quarterfinals before losing in the semifinals to the defending champion, Vlad the Impaler.

In *BattleBots*' Season 3 tournament, Voltronic won two bouts and lost one. In the opening round Voltronic KO'd Greenspan. It defeated Bacchus 25–20 in Round 2, and then was defeated 30–15 by OverKill.

VORTEX

KO'd in the opening round of *BattleBots*' Season 3 middleweight tournament by Musashi, Vortex was built by Matt Hockenheimer of Tipp City, Ohio.

W

WACK-A-TRON

Team Wack-a-tron is run by Dennis Shellhouse of Phenix City, Alabama. In *BattleBots*' Season 3 superheavyweight tournament, Wack-a-tron was KO'd in the opening round by the savage gavel of The Judge.

WACKER

Wacker—built by Bob Schneeveis of Palo Alto, California—advanced to the fifth round of *BattleBots*' Season 3 lightweight tournament. It KO'd Midnight Project, won a 37–8 decision over Eric the Red, outpointed Sparky 34–11, and then was itself KO'd by Ziggo.

THE WACKY COMPASS

Christopher Baron of Roy, Washington, built The Wacky Compass, which KO'd Black Warrior in the opening round of *BattleBots*' Season 3 lightweight tournament before being KO'd itself by Wedge of Doom in Round 2.

WAR MACHINE

A superheavyweight built on wheels and powered by batteries, its primary weapon is a compressed-gas wedge. It is the first battling bot to feature ten-wheel drive. Made of steel and aluminum, War Machine carries three pounds of liquid carbon dioxide.

It was built over the course of eight weeks by head builder Dan Danknick of Orange, California, along with teammates Alvin Buchignani, Roger Brewer, Morgan Boatman, and Mike Herb.

War Machine advanced as far as the semifinals in *BattleBots*' Season 2 superheavyweight competition. It defeated One Tin Soldier, Snake, and, in a close decision, DooAll, before losing in the semis to Diesector.

WEDGEHOG

Wedgehog has a tanklike drive system. Its two wheels are mounted on either side of the robot and are designed to drive independently forward or backward. If one wheel goes backward while the other goes forward, the machine will spin. This feature allows Wedgehog to change direction very rapidly. When both wheels go in the same direction, forward or backward, Wedgehog will travel in a straight line.

Wedgehog was built on a £250 budget. Luckily, the builders had the free use of a friend's workshop, which gave them both creative space and free tools. Before designing Wedgehog, the team checked out the Internet and learned as much as they could about the bots that have come before. They decided to use battery drills as the bot's drive motors. They lacked the machinery and the money to build their own gearboxes, and the drills proved to be an easy-to-install and cheap alternative. Bosch donated the drills to the cause. They were two GSR 9.6 VES-2 battery drills with four batteries and fast chargers. They had two-speed gearing and proved to be easily attached to the bot's axles.

The creative team decided to build Wedgehog's chassis out of wood. This enabled them to alter the design as they went along, drilling holes into the wooden chassis so various parts could be attached in different places. These items were attached with wooden screws. The thinking was that a metal chassis would have been more expensive and less forgiving. One-quarter-inch plywood was used and the eventual chassis was so strong that a grown man could stand on it even after the armor was attached.

After considering all different types of weapons and weapon-powering systems, the team reviewed their budget and decided to use things they already had lying around. They used a pneumatic cylinder and several cans of compressed gas as the actuator.

The switching mechanism was comprised of two 12-volt solenoid valves, a servo, and two microswitches. A couple-and-pulley mechanism was also designed to allow the travel of the piston to be converted into a rotary motion to move a hammer through 180 degrees. To all of this was added a wedge-shaped armor with a curved back and angled sides.

WEDGE OF DOOM

Wedge of Doom was among the class of the field at *BattleBots*' Season 3 lightweight tournament. Built by Tony Buchignani and Team Hazard of Los Angeles, California, the 55-pound Wedge of Doom, which has been competing since 1995, KO'd Lobotomizer, outpointed Dark Matter 39–6, KO'd The Wacky Compass, shut out HammerHead 45–0, barely outpointed Mecha Tentoumushi 23–22, and then lost a 28–17 decision to Gamma Raptor in the semifinals. Wedge of Doom was named the lightweight rookie of the year for the tournament.

Buchignani's day job is as an attorney who specializes in commecial litigation.

WEDGE-O-MATIC

Tim Anderson and the Bot Boys of Granite Bay, California, built Wedge-o-Matic. In *BattleBots*' Season 3 lightweight tournament, it KO'd Undertow in the opening round, outpointed Hobgoblin Ltd. 28–17 in the second round, then was itself KO'd by Reaper in Round 3.

WEED WACKER

Olivier Musy-Verdel and Team Screwed of Menlo Park, California, built Weed Wacker, which was KO'd in the opening round of *BattleBots*' Season 3 lightweight tournament by Death by Monkeys.

WEELIWAKO

In the *Robot Wars* 3 (UK) tournament, Weeliwako defeated Brimhur in the opening round of its heat but was eliminated from the competition by the eventual heat winner, Mace II.

WELD-DOR

At the *Robot Wars* 3 (UK) competition, Weld-Dor defeated the Iron Mask in the opening round of its heat and then was defeated in the second round by King Buxton.

WENDINGO

Wendingo has eight drive motors and features four-wheel drive. It has ten-inch wheels and a seven-inch-tall body, yielding an inch and a half of ground clearance. The body is made mostly of aluminum but the armor is made of a polycarbonate. The primary weapon is the two steel spikes in the front, driven by two motors.

Wendingo's builders are Eric Stoliker, age thirty, of Burbank, California, and Patrick Campbell, age thirty-two, of Stevensen Ranch, California, who have been working in combative robotics for seven years. Patrick is a design engineer, while Eric works as a stage hand for NBC's daytime drama *Days of Our Lives.*

Their robots have fought in Robot Wars '94, Robot Wars '95, Robot

Wars '96, Robot Wars '97, BattleBots '99 Long Beach, BattleBots '99 Las Vegas, BattleBots 2000 San Francisco, and BattleBots 2000 Las Vegas.

In *Robotica*'s Season 1 tournament, Wendingo made it past the preliminary rounds but was ousted by Jawbreaker in the Fight to the Finish.

WHAM! BAM!

Thank you, ma'am. As its name might suggest, Wham! Bam! made only a brief appearance in *BattleBots*' Season 3 lightweight tournament, losing via KO to Half Ass and Random in the opening round. Wham! Bam! was built by Bruce Switzer and Team Peril of Framington, Massachusetts.

WHIRLIGIG

Jeremy Franklin and Jim Smentowski (along with Team Sorceror's Apprentice) of San Rafael, California, built Whirligig, which won a 34–11 decision over Snap Dragon in the opening round of *BattleBots*' Season 3 lightweight tournament, KO'd Recycler in Round 2, and then was itself KO'd by Shrike in Round 3.

WHIRL WEP X25

Michael Winter and the Robot Action League of Walnut Creek, California, built this bot, which KO'd Bulldog in the opening round of *BattleBots*' Season 3 lightweight tournament, outpointed Patriot 26–19 in Round 2, then lost a 23–22 decision to Sublime in Round 3.

WHYACHI

Built by Terry Ewert of Dorchester, Wisconsin, Whyachi—which uses powerful Briggs & Stratton Etek motors—defeated Hammerlock in the opening round of *BattleBots*' Season 3 superheavyweight tournament, then lost a 28–17 decision to Odin II in the second round.

WHYATICA

Whyatica, which recently made its competitive debut at *Robotica II*, was built by thirty-seven-year-old Terry Ewert, a mechanical engineer, Lisa Ewert, and Dale Hammel, all of Dorchester, Wisconsin. The bot has a 24-horsepower power plant, four-wheel drive, and a titanium chassis with a titanium V-plow.

WIDOWMAKER

Kevin and Barbara Lillie of Iowa Park, Texas, built Widowmaker over a five-month period. Built on wheels and battery powered, this bot has a liquid-fuel-powered spinning bar as its primary weapon. A lightweight, Widowmaker is a veteran of *BattleBots.*

WILD WILLY

At the *Robot Wars* 3 (UK) competition, the annoyingly alliterative Wild Willy was eliminated in the opening round of its heat by the even more annoyingly alliterative Flip Flop Fly.

WINDCHILL

Michael Mauldin and Team Toad of Irwin, Pennsylvania, constructed WindChill, which won a 24–21 decision over Phoenix in the opening round of *BattleBots*' Season 3 middleweight tournament, then lost by the identical score to Traxx in Round 2.

THE WITCH

The Witch needed more than a pointy hat and a broom at the *Robot Wars* 3 (UK) competition, where this bot was eliminated in the opening round of its heat by Invertibrat.

THE WIZARD OF SAWZ

Pay no attention to that bot behind the curtain. Some bots run on treads. This bot *is* a pair of treads, with some other parts attached. The treads are so big that the bot operates equally well no matter which side is up.

It has seven carbide-tipped eight-inch circular saw blades as its weapons. Those blades can spin at 5,000 revolutions per minute and the teeth reach speeds in excess of 100 miles per hour. Now that's a satisfying chew!

The Wizard of Sawz was built by Dave Northup, age twenty-five, Seth Dunten, age twenty-six, and Chris Newton, age twenty-six, a group of engineers out of Peterborough, New Hampshire. The bot made its competitive debut at *Robotica II.*

WORLD PEACE

Everybody knows what beauty pageant contestants want more than anything else. World peace, of course. Well, there ain't nothing pretty about this World Peace, a superheavyweight bot that runs on liquid fuel and has a hydraulically powered weapon, a lifting arm and sickle-shaped pecking "beak."

The most noticeable feature of the design are the beverage holders for beer bottles. The bot took eighteen months to build by main builders Greg and Dave Campbell of Fairfield, California, and assistants Stan Kataoka, Ronald Bitondo, Ed Spencer, and Ed Bitondo.

In *BattleBots*' Season 2 superheavyweight competition, World Peace was eliminated by the eventual winner, Diesector. As it turned out, World Peace was as tough on that season's Giant Nut winner as anyone else, leaving Diesector smoking and battered at the end.

At *BattleBots*' Season 3 tournament, World Peace soundly thrashed Junkyard Cog in a 37–8 decision in the opening round before succumbing to Toro by KO in Round 2.

WULFBANE

Bot named after the stuff that is supposed to keep werewolves away—at least according to the old black-and-white Universal movies. This bot was constructed by Michael Jacobson and APSF Robotics Division of Tucson, Arizona.

In *BattleBots*' Season 3 middleweight tournament, Wulfbane KO'd Zombie in an all-Goth opener, then lost a 33–12 decision to Bouncing Betty in Round 2.

X

XAK

Xak was built by Christopher Hannold and Litewav Robotics of Salisbury, North Carolina. Xak stands for eXtreme Ampere Killer, and, at 501 pounds, this is one extreme bot. It is the Haystacks Calhoun of bots.

It is driven by two 24-volt Jensen motors. It features eight-inch lawnmower wheels and 5.4:1 ratio sprockets.

Like the average sumo wrestler, Xak's primary weapon is its own weight. Get into a pushing contest with Xak and you are going to lose, unless you happen to be the reigning Yokozuna.

Xak's armor consists of ½-inch Lexan and a stainless steel frame. It is radio controlled using a Futaba six-channel FM remote and four QD 120-amp speed controllers.

Xak competed in the Robot Battles portion of the annual science fiction convention held in Atlanta, Georgia, called DragonCon, in July 2000. That competition is unique in that you are allowed to use liquid as a weapon.

When beginning construction of Xak, Hannold had the speed controllers, the batteries, the Lexan, sprockets, and the steel frame. The frame, he says, came from the scrap bin of a former employer.

Everything else was leftovers from previous projects. Chris first heard about the Jensen motors while chatting on the *BattleBots* Web site. He then bought them from the C & H sales catalog.

Chris says, "I cut some half-inch plywood to fit inside the frame. Then I cut holes in it to accommodate the wheels and large sprockets. I attached it to the frame with T-shaped mending braces from the local hardware store. The original design was to have the motors mounted on the plywood. However, the 4QD controllers and the batteries took up too much room. I decided to mount the motors on a bracket above the axles. I made the bracket out of three-quarters-by-one-eighth-inch flat black iron. I had planned to make chain tensioners using a quarter-inch steel rod and some nylon spacers but I got lucky and didn't need them. Next time I'll give a shot figuring out the distance between the motor shaft and the axle beforehand."

X TERMINATOR

At the *Robot Wars* 3 (UK) tournament, X Terminator made it to the finals of its heat, defeating Hefty and Judge Shred before being eliminated by heat winner Panic Attack.

Y

YOSHIK

Built by Steven Karp and Team Tekken of Sacramento, California, Yoshik was KO'd in the opening round of *BattleBots*' Season 3 heavyweight tournament by B.O.B.

Z

ZERO

Zero is the first robot to be built by Douan Bingham, age twenty-seven, of Leland, North Carolina. His teammates include his wife, Tonya, and father, Kenny. Douan has a truly great profession, maybe one of the great professions of all time: He is a pro haunted-house builder. He is the Booo-meister! Zero made its competitive debut at *Robotica II.*

ZEUS

At first I thought this bot was named after the king of the Greek gods, but then I came to realize it was actually named after the cross-eyed ham-'n'-egger who once wrestled Hulk Hogan.

At the *Robot Wars* 3 (UK) competition, Zeus was eliminated in the opening round of its heat by the eventual heat winner, Scutter's Revenge.

ZIGGO

A dangerous spinning bot, the 53-pound Ziggo was built in 800 hours by Jonathan, Douglas, and Linda Ridder (Team Ziggy) of San Jose, California. It is battery powered and built on wheels. Its main weapons are ¼-inch tool steel blades powered by 15,000 joules. It has many nicad batteries and works equally well no matter which side is up. Spinning bots can be difficult to steer, but this one has a gyroscope to help maintain its bearings. Jonathan has been building combat robots since 1995.

About the sport, he says, "Robotic combat encourages many good traits—ingenuity, hard work, patience, perseverance. You learn to not take things too serious because all your hard work may get destroyed in a few seconds. Finally, the other competitors are usually very friendly."

At *BattleBots*' Season 1 lightweight tournament, Ziggo defeated Missing Link in the opening round but was eliminated in Round 2 by Das Bot. Ziggo won the big nut at *Battlebots*' Season 2 lightweight tournament. Ziggo not only defeated robots in this tourney, it disected them as well. It sent parts of its foes flying across the BattleBox, defeating Scrap Metal in the opening round, Afterthought in Round 2, and Beta Raptor in the semis. In the finals, Ziggo KO'd Backlash, although this was a brutal bout, with both bots being incapacitated almost simultaneously. The win came at the one minute, twenty-two-second mark.

ZION

Jered Singleton of Reno, Nevada, built Zion, which advanced as far as Round 5 in *BattleBots*' Season 3 middleweight tournament. In that competition, Zion won a 37–8 decision in the opener over Stingray, KO'd Hipopononomus in Round 2, outscored Uplifting Experience 29–16 in Round 3, and KO'd Exodus 2001 in Round 4, then lost a 34–11 decision to Hazard to be eliminated.

ZOMBIE

Zombie was KO'd by Wulfbane in the opening round of *BattleBots*' Season 3 middleweight tournament. It was built by Murray Finegold and Necrobotics of La Mesa, California.

ZUGZWANG

Zugzwang, a lightweight *BattleBots* veteran, was built over a "several-week" period by Chris Conner, Paul La Tour, Marshall Hassan, and Stephen A. La Tour of Wilmette, Illinois. It is built on wheels and battery powered. Its main weapon is a set of three spinning discs.

ZULU

Walter Martines and DreamDroid of Lakewood, California, are responsible for Zulu, which was KO'd in the opening round of *BattleBots*' Season 3 middleweight tournament by Short Order Cheff.

6

TV Personalities

Here's your guide to the guys and gals who host, or have hosted, the robotic combat television programs, who both introduce the contests and offer viewers a guided tour of metallic mayhem—in alphabetical order, of course:

Beiro, Mark

The original idea was to have a comedian or a celebrity announce the BattleBots combatants before the matches and the winners afterward. For one of the early San Francisco shows, Hulk Hogan performed this function.

When *BattleBots* made it to television, it was decided to get a professional boxing ring announcer to do the job, so they got the best, forty-nine-year-old Mark Beiro. Beiro, from Ybor City, Florida, was the regular ring announcer for many years on the USA network's *Tuesday Night Fights,* and is still seen regularly announcing boxing cards shown on ESPN2 and HBO.

Actually, with Beiro, everybody was happy, because they got a pro ring announcer who is a comedian as well. He had no trouble at all keeping a straight face as he read the bots' introductions.

Just check out his Bela Lugosi impression as he introduced Vlad the Impaler! How does Beiro describe his new favorite sport? "The closest thing to it is automobiles in a demolition derby, but this is more high-tech," Beiro says. "I didn't know what to think when I first started doing it, but you have to be impressed with the technology. There's a certain inventiveness there that's very unique. It's like a Hanna-Barbera cartoon gone wild. Hammers come out of nowhere, whirling screwdrivers. And all these things are driven by remote control, by their owners. The split-second reflexes these guys have is absolutely incredible."

BattleBots has raised Beiro's fame a great deal. Beiro says that more people recognize him now than from all of the fights he's announced. "I like it because I'm a ring announcer, basically, in a robot world," Beiro says. "Comedy Central says it's their number-one show ever."

Since an entire season of television shows is shot in a single weekend, and Beiro is the announcer for all of them, he has justifiably earned the name Iron Throat.

Bingham, Traci

Traci Bingham (real name Julie Anne Smith) was born January 13, 1971, in Cambridge, Massachusetts. She is the youngest of seven siblings. She has a diverse heritage, being half Afrrican American, part Cherokee, and part Italian. And all gorgeous.

Like her older siblings before her, Traci attended Harvard. She, however, was more interested in acting. She appeared in sitcoms such as *Cheers, The Fresh Prince of Bel-Air,* and *Married . . . With Children,* a national Coca-Cola commercial, as well as small parts in movies such as *The Firm, Tales From the Crypt Presents: Demon Knight,* and *The Nutty Professor.*

Her big break came in 1996 when she was cast in *Baywatch,* where she starred for two years. She was hired to do post-fight interviews for *BattleBots* for Season 3, and was replaced by Carmen Electra for Season 4.

Charles, Craig

The host of *Robot Wars* in Europe is Craig Charles, an actor-comedian best known for playing the slackerlike Lister for the past ten-plus years in the BBC space sitcom *Red Dwarf.*

He was born in Liverpool on July 11, 1964, at the height of Beatlemania. At the age of twelve he won the *Guardian* newspaper's poetry prize and at fifteen started a band. His interest in music continues to this day with a group he formed called The Eye.

Charles is slated to star in the upcoming *Red Dwarf* movie due out in theaters in 2002. He has also appeared in the film *Business as Usual* and hosted such TV series as *What's That Noise?* and *Space Cadets.*

Danknick, Dan

The head of Team Delta and winner of first-place trophies at *Robot Wars, BattleBots,* and BotBash, Dan is the utmost authority and producer of radio-control equipment for robots. If you are looking to build your first robot, give Team Delta R/C products a look first. Dan has been signed to do on-screen commentary for Seasons 2 and 3 of *Robotica.*

D'Errico, Donna

Baywatch star who performed the post-fight interviews on *BattleBots* during the early days of the show.

Dwyer, Bil

Bil Dwyer has been the host of *BattleBots* since it was first broadcast. He is a stand-up comic who has appeared in movies *(What Planet Are You From?, Ski School,* and *Hugging the Edge),* television *(Ally McBeal, The Larry Sanders Show),* and on the stage *(Wake Up Trenchmont, The Anti-Gravity Treehouse,* and *Velveeta Underground).* He is a former Universal Studios tour guide and his ultimate goal is to be a late-night talk-show host.

Electra, Carmen

The beautiful Carmen Electra got her start in show business as Jenny McCarthy's replacement on MTV's *Singled Out* show. She followed that up with a stint on *Baywatch* and even more courageous duty as Dennis Rodman's real-life wife. She has since left both of those roles for the greener pastures of the *BattleBots* pit area, where she did the post-fight interviews for Season 4, replacing Traci Bingham.

Foley, Mick

In the world of professional wrestling there is no name that conjures up the instant respect among fans as that of the man known as the Master of Mayhem, Mick Foley—a man who gave every ounce of his energy to entertain each and every time he entered the wrestling ring.

Foley, who wrestled under various names, including Cactus Jack, Man-Kind, and Dude Love, is the three-time former WWF world champion, an eight-time tag-team champion and the former Hardcore Champion as well.

He is currently the commissioner of the World Wrestling Federation and still regularly appears on the television shows despite being retired as an active wrestler.

Foley is a multitalented fellow, and has written a pair of best-selling books in addition to his TV work. He has been, since its inception, the host of *Robot Wars: Extreme Warriors.*

Forrester, Philippa

Lovely host of *Robot Wars* in England. She is also the host of the English shows *Tomorrow's World* and *Barking Mad,* in which she has to deal with pets rather than bots. She graduated with honors from London University. She has been the BBC's science girl, hosting *Dreamwheels,* about travel in the future, as well as *Clever Critters, Wild Within, Making Animal Babies* and *Robo Critters.*

Grant, Rebecca

Eye candy for the *Robot Wars: Extreme Warriors* show. Someone should tell her that the word is not pronounced "stragety."

Green, Tim

Former NFL player, a lawyer, and the color commentator who replaced Sean Salisbury as Bil Dwyer's cohost on *BattleBots.*

Mark, Heidi

Heidi Mark handled the post-fight interviews on *BattleBots* before Traci Bingham took over that role. Her film credits include *Rock Star* (2001); *The Judge* (2001), and *Bad As I Wanna Be: The Dennis Rodman Story* (made for TV, 1998). She was featured in the 1998 video *Playboy: Girls of the Internet.* She also made a guest appearance on *Baywatch* in 1995.

Memme, Tanya

Robotica host Tanya Memme was formerly the Fan Cam host on KCBS-TV's *Sports Central* in Los Angeles. She has appeared in *Melrose Place, FX the Series, Saturday Night Live* (segment with Sylvester Stallone), *Another World,* and *Hollywood East.* Her film work includes *Two Hearts One Soul, Below Zero, Lost Soul and Sandals,* all independent productions.

Nye, Bill

Bill Nye, the Science Guy, is a veteran of children's programming on PBS. He does segments on the scientific aspects of robotic combat on *BattleBots.*

Ray, Tanika

One of the hosts of *Robotica,* Tanika Ray is an actor and a dancer with a degree in theater arts from Spelman College. Her previous TV hosting experience came as the cohost of the game show *Wheel of Fortune 2000.* She has also appeared on *Rude Awakenings, NYPD Blue* and *Living Single.*

Salisbury, Sean

Former football player who was Bil Dwyer's original cohost on *BattleBots.* Salisbury now does college football analysis for ESPN.

Sklar, Jason and Randy

Comedic twins, veterans of MTV's *Apartment 2F,* who do post-fight interviews on *BattleBots.*

Zappa, Ahmet

Ahmet Zappa is the son of the legendary leader of the Mothers of Invention rock group, Frank Zappa. Ahmet and his brother, Dweezil, have followed in their father's footsteps and both are professional musicians.

Ahmet has been a popular guest on shows such as *Late Night with Conan O'Brien* and *MTV Live.* Along with his brother, Ahmet was also the host of the USA Network's variety show *Happy Hour.*

7

Tips for New Builders

If you think you can walk into your garage and immediately start to build a battling robot, you couldn't be getting off on a worse foot. Before you can put part A into part B, you have to have a workable design.

The Chassis

In that design, you must first plan out your chassis, which includes the frame and the armor, the outer casing. Part of designing your chassis is determining your bot's appearance.

If you plan to fight your robot in an organized tournament, make sure you know the weight limits that are put upon their competitors. Don't show up for the 125-pound-division tournament with a 135-pound bot. You'll be sent home—or asked to remove ten pounds' worth of parts, a move that could render your bot inanimate.

Materials . . . and Your Budget

The weight of your bot is going to limit everything else you do, especially when determining which materials you are going to use when building your chassis.

In what ways shall you make it strike fear into the metallic hearts of its foes? What will your bot be made of? Steel? Aluminum? Much of this part of your design will be decided for you by your working budget. If you are broke, you probably aren't going to be using bulletproof Lexan or other futuristic-sounding materials. One thing we should make clear right away is that building and fighting a fighting robot *can* be a very expensive hobby. But that does not mean that it *has* to be. Builders have been known to put upwards of $60,000 into their bots, only to see their pride and joy be sliced to pieces in ten seconds by its opponent's buzz saw. Of course, it doesn't have to cost you $60,000 to build and fight a bot. Use parts from an old lawn mower here and an old vacuum cleaner there and you can keep costs low. You might want to avoid using wood, as one poor builder did, especially if you are entering a tournament in which your opponent is allowed to use a flamethrower.

Remember, the heavier your bot is, the more power it will take to make it move. The lighter your bot is, the more likely it will get smashed when rammed in anger.

Making It Move

Next you must determine how you are going to make your robot move. What type of motors will it have? Will it run on separate wheels or on treads like a tank? Will it walk on feet like the stompbots? Your drivetrain must also be determined when you plan your bot's method of movement.

With that planned, you must decide how your robot is to be powered. Will there be batteries? Will a liquid fuel be used? Will your weapon and your wheels be powered by the same or separate sources?

Many robots have an electric-powered drivetrain with a weapon that functions on compressed gas, for example. Now that you know how your robot will work, you must figure out how you are going to work your robot.

Remote Control

A robot's strategy in the combat arena is only as good as the system of control designed by the builder. What remote-control system will you use? How will you control your motor? How will you control your weapon?

Most simple radio-control systems consist of a radio-control receiver attached to an electronic speed controller, which is, in turn, attached to the motor.

According to the Team Delta Web site, "Most competitors use standard hobby radios used to control miniature helicopters and airplanes. This is not intended to be an exhaustive resource on radios but here are the basics. They are available in three flavors, ranging from least to most reliable: AM, FM, and PCM. PCM is a digital version of FM, and although the reception isn't guaranteed, the receiver can detect and reject a garbled packet. Many competitors recommend PCM radios if you can afford it. Some manufacturers are Futaba, Hitec/RCD and Airtronics. Large hobby distributors like Hobby Shack and Tower usually have the best prices. But it appears that Hobby Horse has the best deal on Hitec brand radios. The electronic speed control converts the digital control signals from the radio receiver to high-current signals capable of spinning an electric motor. There are hobby-grade controllers and professional. Both are expensive. Don't believe the '500 amps!'–type rating on the hobby-grade controllers; the reversing models can only handle 40 amps continuously or 80 for a few seconds. And they only work at 12 volts, too. You will need to know the current rating of your motor before you shop for an ESC. Popular manufacturers are Tekin, Vantec and 4QD."

Safety Precautions

If you plan to fight your bot in a tournament, you will need to know what safety precautions the tournament requires and make sure your bot qualifies. You don't want to get all the way to the show and find out that you have inadvertently made an ineligible bot.

"What do you mean, no hand grenades?!"

Weapons

From watching robots fighting, you probably know what sort of weapon you want on your bot. Will it be wedge shaped and ram other bots, trying to get under them and push them into the spikes that line the arena? Or will you have a flipper arm on the front that will get under bots and flip them onto their backs? Will you install a pneumatic hammer that will pound your opponents into oblivion, or will you make your bot a spinner and send other bots flying with a weapon that whips around and around?

Now, here's the tricky part:

When you design your weapon, you must make sure that it will stand up to the rigors of combat. Many is the battling bot that turned into the fleeing scaredy-cat because it lost all of its offense the first time it took a blow from its opponent.

Tools

Now that you have your design and you have purchased your materials, you will need to figure out which tools you will need to build your bot. If you don't plan on putting some money into tools and you don't already have a well-stocked workplace, I suggest you build a lightweight bot.

The heavier the bot, the more likely you are going to need something that you don't have. The bigger bots often require lathes, vertical mills, welding equipment, and band saws, all of which you may not have handy in the garage.

If your budget is limited, and just about everyone's is, check to see if there may be mills and lathes in your area for rent. Don't design something that you can't make. Your ability to build a bot is limited by a number of factors, all of which must be considered when you are designing your bot.

Bang for Your Buck

You must be able to afford all of the materials. A lightweight bot can be built for $500 to 600, but the larger bots will run you four figures. The average heavyweight bot costs between $1,500 and $2,500, but the best bots can cost as much as $5,000 to build.

Technical Ability

You must have the technical ability to build your bot, which means both the expertise and the correct tools, and you must have the spare time to put in all of the necessary labor.

Many builders make the mistake of thinking they can build bots that, once they get down to the nitty-gritty, are far too complicated for them. There is no rule that says that complicated designs are superior to simple ones. In fact, the contrary is often found to be the case. As they say: KISS (Keep it simple, Stupid).

Time

Planning ahead for the amount of time you are going to need to build a bot is one of the toughest things you will have to do. Try to figure out how long it will take if everything goes wrong and then double that time. That is probably a pretty good guess.

Remember, not everyone is good at everything. The person who is good with a motor may not know a hill of beans about remote control. That's why most battling robots are built by teams. Putting their heads together, they have the expertise to build a bot.

It does not take a doctorate degree to build a robot. That isn't to say that your average school kid could do it, but they could if they had a working knowledge of gas- and electric-powered machines.

Work the Web

If you are looking for more tips on how to build a robot, surf the net. From how-to sites to marketplaces for bot parts, the Internet is the preferred method of communication among bot-philes. Check out: Tony's Robotic Resources Page; Steven Nelson's "How to Build a BattleBot"; Jim Smentowski's (Team Nightmare) Tips and FAQ; Puppetmaster Robotics; CoolRobots Builders' Corner; Grayson's Tips; Adam Clark's Building Guide; *Nuts & Volts* magazine. All of these sites can be accessed through links at www.teamdelta.com.

If you are looking to purchase parts for your bot, there is no better place to look today than the Internet. On any given day, a search on the eBay site will reveal many bot parts up for auction. (Looking at it from the other direction, the Internet is also a good place for you to sell some of your extra bot parts, if that is what you want to do.) The days of being limited to the Sears catalog are over.

Good luck!

8

Playing by the Rules

If you are going to build your robot for *BattleBots*, your bot will have to comply with the rules of that particular television show. Those rules, like those of the other competitions, are based on safety and fairness.

The *BattleBots* guidelines, which I'll discuss first, are based on the Society of Robotic Combat's General Competition Regulations, Builders Guide and Teleoperated Category Specifications. The rules are not intended to restrict the creativity of the bot builders. If you have designed a bot that is outside the *BattleBots* guidelines, yet is safe, you may be able to get your design approved.

One Bot per Weight Class

Each builder is restricted to entering only one robot per weight class per competition. Robots may be entered either for duel competition—that is, one-on-one matches that last three minutes or less—or for the Robot Rumble competition, in which many battling bots are in the arena simultaneously for a five-minute free-for-all.

During duel competitions, no robot is ever asked to compete in more than one match within a twenty-minute period. However, in the rumbles, robots are sometimes asked to compete in back-to-back heats.

Therefore, it is necessary for all between-fight repairs to be made in less than twenty minutes. If a robot's next fight is scheduled before adequate repairs from the previous fight can be made, the robot will be forced to forfeit.

Inspections and Tests

All *BattleBots* contestants must pass a technical and safety inspection. At that inspection it will be determined if the robot has adequate covering for its inner parts (armor). That armor must then be removed so that the inside of the bot can be inspected.

The bots are weighed and those that are overweight are sent home.

No wet-cell batteries are allowed. Battery terminals must not be exposed. Fuel tanks must be securely mounted and protected, and, it almost goes without saying, there must be no visible or audible fuel leaks.

There is an operations test that includes a transmitter/receiver test to make sure the bot is receiving adequate signal without interference. Each bot must have an easily accessible master switch and an emergency off switch, and these will be checked to make sure they are operating properly. The weapon is checked to make sure it is safe outside the arena and is functioning properly.

There is a functionality test, which is like a minidriving test, during which the driver must keep the bot under control for a full three minutes. Contest organizers do not want to put bots that behave like blind mice in the BattleBox.

On occasion, a winning bot may be checked after a match in what is called the "impound" area. The time taken to do this inspection does not count against the guaranteed twenty minutes that contestants have between battles.

Separating Signals

Since there are a lot of radio-controlled bots all in the same area, efforts are made to keep those signals from interfering with one another. One antenna clip is provided for each of the frequencies used by the contestants.

Before their transmitter is turned on, the operator must obtain the appropriate clip for his or her frequency and attach it to the transmitter antenna. Clips are obtained from the *BattleBots* Frequency Coordinator.

This clip must remain attached at all times while the transmitter is on. After the transmitter has been turned off, the clip is returned for others to use. Unless otherwise authorized by a *BattleBots* official, use of a transmitter without a frequency clip can result in disqualification.

If something occurs that is not otherwise covered by the rules, then the judges' word goes and all judges' decisions are final.

Those under eighteen are allowed to compete but only if accompanied by at least one qualified adult.

BUILDING TIPS: INSTALLING THE AERIAL

Many battling robots do not perform as well as they might have because of radio interference or lack of control. These problems are often caused by a faulty aerial. To avoid this, check out some of these helpful hints.

- Make sure that your aerial is between 30 and 40 centimeters long. This is long enough to function properly but not too long as to be impractical.
- The best material for the aerial is piano wire.
- The aerial should be fitted to your robot's outer cover. Be sure to slip the aerial through a plastic insulator so that it doesn't make contact with the cover.
- This may sound silly, but accidents do happen: Put something on top of your aerial so that you don't lean down to work on your bot one day and poke yourself in the eyeball.
- Remove the flexible wire from your receiver and replace it with a braided or lapped screened lead, which should be soldered to your receiver. Attach it to your aerial with a push-on terminal. Make sure the braid (or screen) is grounded to your robot's cover. This wire also should be short enough to be practical.
- If you want to further protect your receiver, you may also cover it with a tin or aluminum mesh box.
- When installation is completed, you should run a few tests to see how far your range is. Perform the range check with the robot on its cradle, the transmitter aerial retracted, and the drive wheels off the ground. Operate your bot from up close first, then slowly back away. Make sure that all batteries are fully charged when you do this. You should get good results up to approximately 150 feet.

Determining Winners

As those of you who have watched *BattleBots* on TV know, duel matches are started with the two bots on opposite ends of the arena. At the start, the bots must be motionless, with all rotary weapons spun down. Internal-combustion engines may be running at idle.

It is important to note that "playing possum" is a strategy that is frowned upon. Bots that do not move are apt to be declared incapacitated and eliminated from the contest. Besides, passive bots are unpopular.

If a duel contest goes the full three minutes, the judges decide the winner. Three judges award 15 points each (a total of 45 points) as follows: Five points are split between the two bots, with the majority of points going to the bot that the judge determines to be the most aggressive. Another five points are split between the two bots, with the majority of points going to the bot that causes the most damage to the other bot. The remaining five points are split between the two bots, with the majority of points going to the bot that employs and executes a better strategy. The bot with the most points wins.

The winners of the rumbles are determined by audience applause, and experience has proven that the crowd loves aggressiveness. Only those bots that are still operating at the end of the rumble are eligible to win.

Pit Crew Restrictions

The number of pit crew members a team is allowed depends on the weight of the bot. The heavier the bot, the more personnel you are allowed. Five crew members are allowed for the superheavyweight division, four for the heavyweight division, three for the middleweights, and two for the lightweights.

Crew members are required to wear Pit Passes at all times during a competition. Crew members must provide a passport-size photo for identification purposes to be laminated into the Pit Pass. Pit Passes are non-transferable or exchangeable. Contestants should practice common sense and good sportsmanship at all times.

While in the pit, your bot's wheels must be lifted off the ground so that they are suspended in air. This is to prevent runaway bots.

Any bot that uses pneumatic/hydraulic systems is not allowed to have those systems pressurized while in the pit area. Pressurization may take place in the BattleBox or in a designated testing area.

Sharp objects and edges on your bot must be kept covered while the bot is in the pit area. Fueling and refueling is also prohibited in the pits. There is a designated safe area for gassing up.

No welding or grinding is allowed in the pits. Again, there are areas set aside for precisely those activities. The welding and grinding ban includes the sanding of plastic and fiber-composite materials—such as those (Lexan, fiberglass, Kevlar) that are commonly used for bulletproofing and bot armor.

BattleBots recommends that all bots use Pulse Code Modulation (PCM) radio transmitters and receivers. Frequency Modulated systems (FM) are technically allowed but frowned upon because of their potential for causing radio interference.

AM radios are prohibited, and all bots must have a "robust fail-safe feature" to prevent loss of control in case radio frequency is lost.

Weight Classes

BattleBots divides its competing robots into four weight classes, each of which has slightly different limits for wheeled and nonwheeled bots. Bots that walk on legs are allowed to be heavier.

The superheavyweight division accepts wheeled bots that weigh between 211 and 326 pounds, and nonwheeled bots that weigh between 316 and 489 pounds. Heavyweights may weigh 116–211 pounds if on wheels and 144–316 if nonwheeled.

Middleweight bots may weigh between 59 and 116 pounds if wheeled and between 88 and 174 pounds if nonwheeled.

The lightweight division accepts wheeled bots between 25 and 59 pounds and nonwheeled bots between 25 and 88 pounds.

You should weigh your bot when it is "wet," that is, filled with any fluids that it might contain when it is fighting, such as hydraulic fluid, gasoline, and compressed CO_2.

If your bot is built in a modular fashion and can battle in various configurations—usually with different weapons—the bot must weigh in under the limit in all of its configurations.

There are no limits on the dimensional size of a robot. As long as it makes the weight limit it is cool. Well, there is one restriction: The bot must fit through the door into the BattleBox since assembly inside the Box is forbidden. That door is eight feet by eight feet, so just about all bots fit.

You may use any material you like to build your bot, and you may use as many different materials as you like. Forbidden materials would include only those that are intrinsically dangerous, such as radioactive materials or asbestos.

Bots may not fly, although they are allowed to hover. Bots are only allowed to go airborne if they just got knocked to kingdom come by their opponent's weapon.

BUILDER'S TIP: PREVENTING RADIO INTERFERENCE

If your electric-motor drive is having problems from radio interference due to noise from the brush armature contact, here are some tips:

- Check the brushes, especially if you are using a secondhand motor.
- Make sure nothing is missing or damaged on the armature.
- Check the motor to make sure no iron filings have been drawn into the motor by the motor's two large magnets. Future problems with filings and other metal bits getting into the motor can be avoided by covering all of the air vents on your motor with masking tape.

Treads, Wheels, and Walkers

The A–Z guide to the bots in Chapter 5 makes reference to "wheeled" bots. A wheeled bot is any that has a direct, continuous, linear relationship between a rotary actuator and its translational displacement.

What is and isn't a walking bot has been a matter of debate for years—and the distinction is important because, as we have seen, *BattleBots* rules give walking bots a weight advantage.

Here's the official definition of a stompbot, or walking robot: "Any robot that uses linear actuators exclusively without any wheels. Any robot that uses a combination of linear actuators and rotary actuators that are driven in an oscillatory manner, that is not continuous."

BattleBots will sometimes approve rotary-actuator walkers that rely on "cams" for walking motion, but that approval must take place before the bot will be allowed to compete in a tournament.

It is allowable to use a rotary actuator to drive a stompbot provided the motion is not continuous. The point here is to avoid bots that have spokes protruding through wheel "legs."

It is allowable to have undriven wheel-type support devices (such as ball casters) on nonwheeled bots, as long as they do not support more than 50 percent of the bot's weight at any time.

Bots that have treaded wheels, like those of a battle tank, are considered wheeled bots in terms of the rules.

Jumping

Although there hasn't been much of it, jumping is allowed according to *BattleBots* rules. The only limits are that bots with hop, skip, and jump capabilities must not be taller than six feet at the apogee of their leap.

And the bot must be designed so that it does not do damage to the floor of the arena when it lands.

Fuel

The *BattleBots* rule makers measure the danger of a bot in terms of the potential energy that is stored in its power sources. Therefore, for everyone's safety, there are limits to just how powerful a bot can be.

All batteries must have a construction that utilizes immobilized electrolytes only. Common types of these batteries are nickel-cadmium (nicad) and sealed lead-acid gel cells. This is because bots get flipped a great deal and all batteries must be safe to use no matter what position they are in. There are no voltage limitations, but teams that operate high-voltage bots are strenuously urged to use maximum safety precautions.

It only takes one french-fried roboteer to ruin everyone's day!

Compressed gas is used to power many of the bots' weapon systems. There is no limit to the amount of compressed gas that can be stored on a bot, but its pressurization is limited to 2,500 pounds per square inch for a filament-wound bottle. If you are not using one—or if you are using liquefied gas—the pressurization is limited to 1,000 pounds per square inch.

If a bot is using liquid fuel, such as gasoline or alcohol, all fuel lines must be protected by metallic braid, and all ends must be clamped.

Remote-mounted fuel tanks are required to be impact resistant and reasonably protected within the bot's body. In case the bot is flipped over, fuel vent systems must be designed such that they will not continuously leak fuel.

Superheavyweights are allowed to carry 21 ounces of fuel. Heavyweights and middleweights can carry no more than 15 ounces, and the limit for lightweights is 12 ounces.

BUILDER'S TIP:
STAY COOL

Electric motors can get hot in the heat of combat, so place capacitors across the terminals of electric motors to prevent voltage spikes from interfering with the control electronics.

Drive Systems

There is no limit to the size or power of the electric motors that can be used. Hydraulic systems have no limits on the quantity of fluid, but pressurization is limited to 5,000 pounds per square inch.

If an accumulator or other volume-building device is used, maximum pressure is restricted to 1,200 pounds per square inch.

Pneumatic rotary drives that use pressurized gas are limited to 2,500 psi. If liquefied gas or non-filament-wound bottles are used, the limit drops to 1,000 psi.

An internal-combustion system may rate an unlimited number of horsepower, but all such engines must have fail-safe shut-off systems in case of emergency. Those engines must also have a throttle return spring that returns the engine to idle when power to the control actuator is lost. The engine must also be designed to return to idle or shut off completely if radio contact is lost with the driver. It is necessary to have a centrifugal clutch or other mechanism that won't allow drivetrain motion or weapon movement while the engine is idling.

Weapons

The following weapons are forbidden by *BattleBots:* any weapon that uses electricity (stun gun); any radio-frequency-jamming equipment; any use of liquid as a weapon; bright lights; any weapon that obstructs the vision (smoke bombs, etc.); any weapon that fires an untethered projectile (tethered projectiles are allowed with tethers up to ten feet); any weapon that uses heat or cold to injure an opponent (flamethrower, fire extinguisher).

Magnets are allowed but are discouraged because of their tendency to cause radio interference. Bots that use magnets must be inspected to prove they do not interfere with radio signals.

While there is no official limit to the number of weapons a bot may have, the allowed overall weight of the bot is the determining factor. The more weapons the bot has, the heavier it will be.

All spinning weapons (and spinning bots, for that matter) must stop spinning within thirty seconds of being turned off, or at least slowed to a point where they can be safely handled by a person without protective equipment.

Robot Wars Rules

Robots Wars bots are allowed to have interchangeable body panels, but those panels must be able to be switched in less than twenty minutes, and no panel switching is allowed during a fight.

All parts that may need attention between matches must be easily accessible. All bots must have a cradle to sit in, so that the wheels are supported off the ground and its drive can be demonstrated without the bot moving.

All bots are inspected for safety and reliability before being allowed in the arena. Bots with wheels, treads, or "feet" are allowed. Flying bots that use exposed rotating aerofoils are forbidden.

In *Robot Wars,* heavyweight bots are all bots between 100 and 220 pounds (which is the equivalent of 45 to 100 kilograms in the metric system). As is true in *BattleBots,* walking bots get a weight advantage. In this case, stompbots are allowed to compete in one weight class below their actual weight. Bots are weighed in the configuration they will be in during battle and without fuel.

Bots may not be longer that 6.6 feet or wider than 4.6 feet. Stompbots may be larger but require prior approval from tournament organizers. There is no limit to a bot's height.

Power

Bots may be powered by electric, internal combustion, hydraulic, or pneumatic systems. The bot and its weapons do not have to be powered by the same sort of system.

Emergency off switches are required, removing all power from the bot. If more than one switch is needed to turn off the bot, those switches must be easily accessible and in close proximity to one another. The switches can be covered but the cover must be easily removed without tools.

Electric motors must have working voltages that do not exceed 36 volts of direct current, or main voltage. All connections must be adequately insulated. Cables must be routed. Batteries must be sealed and must not contain free-flowing liquid.

For internal-combustion engines, fuel capacity is limited to 17 fluid ounces. The tank must be made of plastic or, if metal, have a plastic "pop off" seal. The tank must be puncture-proof.

A return spring must be fitted to the throttle of all internal-combustion engines to return the throttle to "idle" or "off" in the case of servo breakage or failure.

For hydraulic engines, pressure is limited to 3,000 pounds per square inch, and the storage tanks for the hydraulic fluid must be made of an appropriate material.

With pneumatic systems, pressure is limited to 1,000 pounds per square inch. All gases in pneumatic systems must be inert or nonflammable—for example, air, carbon dioxide (CO_2), argon (Ar), or nitrogen (N_2).

Weapons

Robot Wars does not allow pyrotechnics or radio-jamming weapons. Only the house bots—bots that patrol the arena looking to fry the losers—are allowed to shoot flame.

Untethered projectiles are not allowed and tethered projectiles must not have a tether longer than eight feet. Spinning weapons must not spin faster than their manufacturer's recommendations.

Hardened steel blades that run the risk of shattering are not allowed. Any sharp blade may not be longer than eight inches. All sharp edges must be covered when the bot is not in the arena.

If the weapon involves a swinging arm or other part, that part must be locked into place and immobile when the bot is not in the arena. Lasers are allowed up to Class 2.

Radio Control

All *Robot Wars* bots must use VHF frequency band 75Mhz. A quality set (that is, one that uses interchangeable crystals) must be used. All dangerous operating circuits must have a "fail-safe" device fitted to each circuit.

At least three pairs of transmitter and receiver crystals of different frequencies must be available for each radio-control setup involved in running the bot. If two separate transmitter/receiver sets are used, there must be six different pairs of crystals available.

So there you have it. The rules. Every one of them should be kept in mind before you go into the workshop to start to design and build your own bot. There's no sadder face than that of the builder of a bot who has proudly brought his machine to the meet, only to be sent home by the judges.

Glossary
A Guide to the Lingo of Robotic Combat

Autonomous A robot that operates without a driver. An autonomous robot would need to be preprogrammed through a computer. A radio-controlled robot, such as the great majority of combative robots, is not autonomous.

C.A.D. Computer-aided design.

ClampBot A robot whose primary weapon are pincers.

StompBot A bot that walks on feet.

Controller Mechanism by which the driver controls the bot through radio waves.

ESC Electronic speed control. Controls the speed at which the wheels turn.

Flipper Weapon designed to get under the opposing bot, then snap upward, flipping the foe onto its back.

Ground clearance The room between the bottom of a robot and the floor. Ground clearance comes in handy on the *Robotica* TV show, where obstacles must sometimes be climbed over.

HammerBot Robot that features a hammering tool, sometimes an ax, as its primary weapon.

Invertible Able to function with either side up.

IRPD Infrared proximity detector. As the popularity of combat among autonomous robots continues to grow, expect to hear more talk of these, which use reflecting heat to judge a robot's proximity to other objects.

NiMH Nickel Metal Hydride.

NPC National Power Chair, makers of the finest in wheelchair motors.

Pits An area in close proximity to the combat arena in which teams prepare their bots before bouts and repair bots between them.

Polycarbonate A durable plastic.

Self-righting The skill some robots possess of being able to turn themselves right-side up in case they get turned upside down.

Servo Motor with internal location sensor and gear reduction.

SLA A good battery. It means sealed lead acid.

SpinBot Bot that goes round and round or has a weapon that does the same.

Ti Short for titanium, both here and on the golf course.

Turning circle Space needed to turn a bot 180 degrees.

WedgeBot Bot that's shaped like a doorstop, designed to get under and control an opponent with its shape.

Index

NOTE: For names of individual robots and team members, see Chapter 5, Encyclopedia of Battling Robots.

About the Author

Michael Benson is the former editor of the *Military Technical Journal* and *Stock Car Spectacular* magazines. He is the author of eighteen books, including *Dale Earnhardt* (Chelsea House, 1996), *Monster Trucks* (Smithmark, 1997), and *Crashes and Collisions* (Chelsea House, 1998). Originally from Rochester, New York, he is a graduate of Hofstra University. He lives in Brooklyn, New York, with his wife and two children.